SAFE DRINKING WATER

The Impact of Chemicals On A Limited Resource

113 N. Henry Street, Alexandria, VA 22314 (703) 683-5213

SAFE DRINKING WATER

The Impact of Chemicals On A Limited Resource

Rip G. Rice, PhD
Editor for the Drinking Water Research Foundation

Library of Congress Cataloging in Publication Data
Main entry under title:

Safe Drinking Water.
Bibliography: p.
Includes index.
1. Drinking water—United States—Contamination.
2. Chemicals—Environmental aspects—United States.
3. Water—Pollution—United States. I. Rice, Rip G.
RA592.A1S24 1984 363.6′1 84-25105
ISBN 0-9614032-0-9

Drinking Water Research Foundation
113 North Henry Street, Alexandria, Virginia 22314

PRINTED IN THE UNITED STATES OF AMERICA

PREFACE

A review of the nation's news coverages serves as a ready reminder that drinking water safety is more than a regional or local concern. In recent times, the print media alone has drawn attention to barium, bacteria, heavy metals, and increasingly organic contaminants, in public water supplies located in Florida, Rhode Island, Texas, Oregon, Illinois, Minnesota, North Carolina, Michigan and California, to name a few.

In an effort to address one of the major issues confronting the future of the nation's drinking water supplies, chemical contamination, the Drinking Water Research Foundation and the American Chemical Society presented the symposium, "Safe Drinking Water: the Impact of Chemicals on a Limited Resource." To add balance to the total presentation, two papers were included that were not part of the symposium.

Many questions as to the public significance of hundreds of organic chemicals known to be present in the national drinking water supply are waiting to be answered. In some areas of the country, acid rain-induced alterations of the natural leaching process represent an unexplored potential source of toxic pollutants. Finding workable ways to clean up the water supply will be an ongoing task.

Addressing these questions, as well as investigating how other countries are responding to these problems, the alternate sources available, such as bottled water, and point-of-use devices, the presentors in this symposium have attempted to explain the problems, situation, and alternatives.

As progress is made in one area, setbacks will occur in another. As we eliminate problems through chemical technology, we often create others, such as contamination of our waters.

While all the situations, problems, and alternatives are not discussed in these proceedings, it is hoped that some attention will be brought to the public, government, and private sectors so that future work will be done to assure the nation of safe drinking water resources.

William F. Deal, CAE
Executive Vice President
Drinking Water Research Foundation

EDITORIAL ACKNOWLEDGMENT

Dr. Rip G. Rice received his PhD degree in Organic Chemistry from the University of Maryland, and has more than 36 years of varied experience in the federal government, aerospace, the chemical industry, and as a private consultant in government liaison, environmental engineering, and water and wastewater treatment.

In 1973, Dr. Rice co-founded the International Ozone Association, is a member of its Board of Directors, and has served as its President during 1982 and 1983. During the period 1977–1980, he was a member of a consulting team of specialists funded by the U.S. Environmental Protection Agency to survey and analyze European and Canadian water treatment practices involving the uses of ozone, chlorine dioxide, granular activated carbon, and biological treatment processes. Recently, Dr. Rice conducted an extensive literature review of the oxidation products formed when drinking water is treated with various oxidizing agents, for EPA's Office of Drinking Water.

Dr. Rice is coeditor of the series of books entitled, "Handbook of Ozone Technology and Applications," and coauthor of the book, "Biological Activated Carbon," both published by Ann Arbor Science Publishers, Inc. In addition, Dr. Rice has edited 10 monographs and proceedings for the International Ozone Association.

A member of the AWWA, WPCF, Sigma Xi, IOA, AiChE, the International Bottled Water Association, and the Technology Transfer Society, Dr. Rice recently has been appointed to the Joint U.S./USSR Work Group on the Prevention of Water Pollution from Industrial and Municipal Sources, directed by the U.S. Environmental Protection Agency.

* * * * * * * *

The editor is indebted to Miss Denice K. Cassady, whose outstanding editorial and manuscript preparation services have made this opus possible. Her constantly cheerful demeanor in the face of much drudgery was an inspiration, and was very much appreciated.

TABLE OF CONTENTS

Part I—NATIONAL AND INTERNATIONAL PERSPECTIVES

1. Chemicals and Safe Drinking Water: National and International Perspectives 1
   Robert A. Neal, Ph.D.

2. Drinking Water: A Global Victual 9
   J. R. Hickman, B. Pharm, M.S.

Part II—SOURCES & DISTRIBUTION OF SAFE DRINKING WATER

3. Public Drinking Water and Chemicals 21
   James F. Manwaring, P.E.

4. Bottled Water: An Alternative Source of Safe Drinking Water 33
   Jerry T. Hutton, Ph.D.

5. Overview of Point-Of-Use Water Treatment Technology 43
   P. Regunathan, Ph.D.

Part III—CHEMICALS IN DRINKING WATER

6. An Effective Alternative to Official Regulation of Indirect Additives to Drinking Water 63
   Nina I. McClelland, Ph.D.

7. The Water Treatment Chemicals Codex 83
   Robert A. Rehwoldt, Ph.D.

8. Disinfectant Chemistry In Drinking Water—Overview of Impacts on Drinking Water Quality 87
   A. A. Stevens, L. Moore, R. C. Dressman, and D. R. Seeger

9. By-Products of Chlorination: Specific Compounds And Their Relationship to Total Organic Halogen 109
   Daniel L. Norwood, G. P. Thompson, J. J. St. Aubin, D. S. Millington, R. F. Christman, and J. D. Johnson

10. Ozone for Drinking Water Treatment—Evolution and Present Status 123
Rip G. Rice, Ph.D.

11. The Occurrence of Contamination In Drinking Water From Groundwater Sources 161
Hugh F. Hanson, P.E.

Part IV—MONITORING AND ANALYSIS

12. Improved Monitoring Techniques to Assess Groundwater Quality Near Sources of Contamination 167
Glenn E. Schweitzer

13. NBS Environmental Standard Reference Materials for Use In Validating Water Analysis 177
R. Alvarez

Part V—REGULATORY ASPECTS

14. Regulation of Contaminants In Drinking Water 183
Joseph A. Cotruvo, Ph.D.

15. Federal Protection of Groundwater 197
Timothy L. Harker, Esquire

16. The World Health Organization and Guidelines and European Economic Community Directives 201
J. R. Hickman, B. Pharm, M.S.

17. Regulatory Flexibility and Consumer Options Under the Safe Drinking Water Act 209
Timothy L. Harker, Esquire

18. Structure and Regulation of the European Bottled Water Industry 223
Rip G. Rice, Ph.D., and G. Wade Miller

19. Congressional Initiatives 257
Honorable Robert T. Stafford

20. Conference Summation 263
Timothy L. Harker, Esquire

CHAPTER 1

CHEMICALS AND SAFE DRINKING WATER: NATIONAL AND INTERNATIONAL PERSPECTIVE

Robert A. Neal, Ph.D.

Chemical Industry Institute of Toxicology
Research Triangle Park, NC 27709

Until quite recently, "pure" drinking water was judged by the absence of mud, taste, odor, color, fecal coliforms, total dissolved solids and a few other factors. However, the realization that drinking water is contaminated by a myriad of inorganic and, particularly, organic chemicals has changed the way we judge its purity. This change in the way we view our drinking water has been a natural outgrowth of the drastic change in the nature of the diseases that account for mortality and morbidity in our society. Only a few decades ago, infectious diseases and diseases caused by parasites were the major causes of morbidity and mortality. Now cardiovascular diseases and cancer are the prevailing threats to human health. The determination of the role of life-style and chemical contaminants in our environment, including drinking water, in the etiology of these diseases is now a major activity of governments, scientists and the medical profession. The purpose of this paper is to examine, in a general sense, the issue of chemical contamination of drinking water sources and, to the degree possible, the implications of that contamination to human health.

Certain human activities provide important sources of chemical contamination of drinking water. Among these are waste disposal from manufacturing and mining, the use of pesticides and fertilizers in agriculture, the disposal of domestic wastes including sewage, wastes from animal feedlots and highway deicing materials.

The decay products of vegetation are an important source of chemical contamination of drinking water largely unrelated to human activities. Various chemicals used for treatment and disinfection of water react with these decay products of vegetation to produce some compounds of potential concern for human health. An example is the production of chloroform and other trihalomethanes as a result of the use of chlorine in the treatment and disinfection of water supplies containing the vegetation decay products.

A number of the inorganic chemicals which are detected in water supplies are also of natural mineral origin. Other chemicals occur as contaminants in drinking water as a result of materials used in the water distribution systems. An example is the occurrence of polyaromatic hydrocarbons in drinking water as a result of the use of tar in coating pipes used in the distribution system. Radionuclide contamination in drinking water can be of both natural origin or a result of human waste disposal practices. Those of natural origin currently are of greatest concern.

Chemicals found as contaminants in drinking waters can be grouped conveniently into three classes: inorganic, organic and radionuclides. A number of inorganic compounds are found as contaminants in drinking water. As noted previously, some of these are of natural origin. One of these, fluoride, often is purposely added to water supplies as a prophylactic against dental caries. Some of these inorganic contaminants of drinking water are of concern as regards potential adverse human health effects. These are:

- Arsenic (cancer)
- Cadmium (kidney damage)
- Chromium (cancer)
- Cyanide (acute toxicity)
- Fluoride (skeletal fluorosis)
- Lead (nervous system)
- Mercury (nervous system)
- Nitrate (methemoglobinemia)
- Selenium (gastrointestinal)

A number of governments and international bodies have recommended maximum levels for these inorganics in drinking water.

Inorganic contaminants other than those noted above for which there is some, but much less concern for human health, are aluminum, barium, asbestos, beryllium, nickel and silver. A few governments have set maximum concentration levels for some of these inorganic compounds.

Other inorganic chemicals contribute undesirable aesthetic or organoleptic properties to drinking water. These are copper, iron, magnesium, manganese and zinc, and levels consistent with good water quality also have been established for some of these.

In the past decade several hundred different organic chemicals have been identified as contaminants of drinking water. However, those so far identified represent but a small fraction of the total organic matter present in drinking water. The majority of the organic compounds present in drinking water consist of nonvolatile, water-soluble substances, a substantial fraction of which have molecular weights in excess of 1,000. The identification of the structures of most of these nonvolatile compounds will be exceedingly difficult.

In a conventional sense, it will be necessary to know the structure of these compounds before it is possible to begin an assessment of their potential toxicity to humans consuming the water. However, because of the enormous number of compounds normally present and the difficulty in identifying the structure of the majority of the compounds, some consideration is being given to carrying out animal studies on a representative mixture of the organic compounds present in a drinking water supply, then basing the assessment of the potential toxicity to humans consuming the water on the results obtained in animals using the mixture.

In the meantime, attempts to control the levels of potentially harmful organic contaminants in drinking water is based on examination of drinking water supplies for specific compounds or classes of compounds known to be potentially toxic to humans. The classes of compounds of particular interest include the halogenated alkanes and alkenes, the polycyclic aromatic hydrocarbons, benzene, and various alkyl-substituted benzenes, chlorophenols, chlorobenzenes, nitrosamines, pesticides, aromatic amines, aromatic nitro compounds, polychlorinated biphenyls and the phthalate esters.

The chlorinated alkanes and alkenes of greatest interest based on their occurrence and evidence of toxicity in experimental animals are:

- Chloroform (and other trihalomethanes)
- Trichloroethylene
- Tetrachloroethylene
- Carbon tetrachloride
- 1,1,1-Trichloroethane
- 1,1-Dichloroethylene
- 1,2-Dichloroethane
- 1,2-Dichloroethylene
- Trichloroethylene
- Methylene chloride
- Vinyl chloride

The pesticides of most interest based both on occurrence and/or evidence of toxicity or potential toxicity to humans are:

- DDT
- Aldrin
- Dieldrin
- Heptachlor
- Heptachlorepoxide
- 2,4-D
- 2,4,5-T
- Chlordane
- Lindane
- Methoxychlor

- Benzene hexachloride
- Triazine herbicides
- Hexachlorbenzene
- Endrin
- Toxaphene
- Aldicarb

In addition to benzene itself, the alkylbenzenes, halogenated benzenes and chlorinated phenols of greatest interest are:

- Toluene
- Ethylbenzene
- m-Xylene
- 1,2-Dichlorobenzene
- 1,4-Dichlorobenzene
- Trichlorobenzenes
- Bromobenzene
- o-Chlorotoluene
- Pentachlorophenol
- 2,4,6-Trichlorophenol.

Guidelines suggesting maximum concentrations based on potential health effects have been established, or soon will be established by some governments and international bodies for a few of these compounds. Most notable are:

- the Trihalomethanes
- Endrin
- Lindane
- Methoxychlor
- Toxaphene
- 2,4-D
- 2,4,5-T
- DDT
- Chlordane
- Hexachlorobenzene
- 1,2-Dichloroethane
- 1,1-Dichloroethylene
- Pentachlorophenol
- Tetrachloroethylene
- Trichloroethylene
- Vinyl chloride
- 2,4,6-Trichlorophenol

The radionuclides which occur as contaminants of drinking water which are of greatest concern for human health are radium-226 and -228, uranium and radon. All of these are of natural origin and are a health concern because of the potential for radiation-induced cancer.

Although large numbers of potentially toxic organic chemicals have been detected as contaminants of drinking water, in fact, only a small fraction of the organic chemicals detected as contaminants in drinking water have been examined by available means for their potential to cause toxic effects in humans. It is logical to assume that when properly tested, some of these compounds will show evidence of a potential to cause toxic effects in humans consuming the water. However, in considering the potential of contaminated drinking water to cause adverse health effects in humans, it is important to keep in mind that in terms of the number of people affected, microbial contamination of water is of far greater concern than chemicals. The adverse health effects experienced when consuming water excessively contaminated with microorganisms which are pathogenic to humans usually occur shortly after ingestion and if they do not cause mortality, usually are of short duration. In contrast, the potential human diseases caused by some chemical contaminants in water are largely believed to be delayed effects which occur as a result of consuming the contaminanted water over extended periods of time.

The greatest public concern is for the potential of chemical contaminants in drinking water to cause cancer; this concern also is reflected in the activities of governments. However, knowledgeable scientists and health professionals believe that some attention also should be paid to the ability of chemical contaminants in drinking water to adversely affect other organ systems or biological processes which do not easily repair themselves or in which the effects are not easily reversible. These include effects on the fetus, germ cells, the reproductive systems, and the nervous system.

In evaluating the human health effects which may occur upon exposure to chemicals occurring as contaminants in drinking water, there is an implicit search for absolute answers on the part of most governments, regulatory agencies and the public. Witness the increasing uses of mathematical models to carry out quantitative risk assessments for human cancer incidence using data from studies carried out in experimental animals with compounds present as contaminants in drinking water. Data obtained from these mathematical models generally would not be accepted for publication by refereed journals publishing biological data because currently there is no experimental means for verifying their accuracy. Yet these numbers are largely perceived by the public to be a precise statement of the risk. An important but difficult task for the informed scientific and medical community is to educate the public, government officials, decision-makers within regulatory agencies and the less informed medical and scientific community that with the technology currently available we are not able to make precise judgements about the potential health risks inherent in the consumption of chemically contaminated drinking water.

In evaluating the potential for human toxicity of chemical contamination of drinking water, three methodologies are available (Neal, 1982). These are epidemiological investigations in populations consuming the contaminated water, "short-term" tests for the presence of mutagenic and potentially carcinogenic chemicals and studies in experimental animals.

To date, epidemiological studies have revealed a few instances of acute or subacute toxicity resulting from humans consuming water contaminated with chemicals. Most of these have involved contamination with inorganic chemicals. Notably, these have involved gastrointestinal disturbances as a result of improperly designed, improperly operated or malfunctioning fluoridation equipment, gastrointestinal disturbances as a result of leaching copper from distribution systems, and, on rare occasions, nitrate-induced methemoglobinemia (Craun, 1981). Additional epidemiological studies have failed to show a consistent relationship between cancer and chemical contamination of water (NAS, 1980). Also, a causal relationship between water hardness and heart disease has not been borne out by epidemiological investigation (NAS, 1980). Finally, a relationship between water hardness or lead, nitrate, sodium or cadmium contamination of water and hypertension has been suggested. However, the evidence from properly conducted epidemiological studies has been inconclusive regarding these relationships (Folsom and Prineas, 1982).

Thus, except for an occasional outbreak of acute toxicity largely resulting from defects in water treatment or problems with the distribution system, epidemiological studies have not revealed definitive evidence of human illness as a result of chemical contamination of water.

A number of short-term tests for mutagencity or mammalian cell transformation have been used to detect mutagens and putative carcinogens in drinking water supplies. However, a major shortcoming of these short-term tests is that they will detect only potential mutagens or putative human carcinogens. No short-term tests have yet been developed to determine the potential of the chemical contaminants to produce toxic effects in the fetus, the reproductive systems, the nervous systems or other biological processes in mammals. In addition, the results of short-term tests for putative human carcinogens correlate only approximately with the results of cancer studies in experimental animals. Also, short-term tests for putative carcinogens do not detect promoters of cancer, nor do they provide reliable data on the potency of initiators of cancer in experimental animals and, by inference, in humans. Thus, it is not possible, at present, to use data generated by applying short-term tests to chemically contaminated drinking water to determine with sufficient confidence the human health risk which may occur from consumption of that water.

The results of animal bioassays, conducted largely in rodents, have been and will continue to be the major source of data for estimating the potential risks to humans posed by consuming chemically contaminated water. Although there are limitations in the ability of animal tests to predict the potential for human disease to result from consumption of chemically contaminated water, it is the best methodology currently available.

As noted previously, a major difficulty in judging the potential risks to humans of exposure to chemically contaminated drinking water is that any toxicity that occurs would be the result of exposure to a complex mixture of chemicals. The technical aspects of animal tests using mixtures of chemicals are still under development. Thus, systematic animal studies of the mixtures of chemicals present in a representative drinking water supply have not yet been carried out. Up to the present time, we have been limited to estimating the approximate potential risks to humans which are implied by the presence in drinking water of individual compounds of known toxicity based largely on the results of animal tests.

SUMMARY

A large number of microorganisms and chemicals occur as contaminants of water intended for human consumption. The human health implications of microbial contamination of water are far more severe than those attendant with chemical contamination. The contamination of drinking water by microorganisms pathogenic to humans can be controlled largely by the disinfection techniques currently available. The assessment and control of the potential health effects attendant with chemical contamination of water are much more difficult. However, because of its primary importance, the control of microbiological contaminants should never be compromised in order to control the contamination of the water with chemicals.

Epidemiological surveys of populations consuming chemically contaminated water so far have to show a consistent relationship between the chemical contamination and alternations in the incidence of cancer, heart disease and hypertension. However, because of the inherent insensitivity of epidemiology, it cannot be assumed that there is no risk to human health from these contaminants. So-called "short-term" tests have revealed the presence of putative human carcinogens in finished drinking water. However, because of some lack of correlation between the results of short-term tests and the induction of cancer in experimental animals exposed to a variety of chemicals and the inability of short-term tests to actually predict carcinogenic potency, the results of these tests cannot be taken as evidence of the presence or absence of a cancer risk to humans consuming the water.

A few compounds which produce cancer in humans and a large number which produce cancer in experimental animals have been detected in finished drinking water. This information has been used to propose limits on the concentrations of these compounds which should be allowed in finished drinking water. However, there is no information on the potential human toxicity of the majority of the organic compounds present in drinking water. Since it will not be possible nor desirable to test each of these compounds for possible toxic effects in humans, consideration is being given to carrying out animal tests with representative mixtures of the chemicals present in drinking water supplies. This is a toxicologically sound approach, since the testing of the mixture automatically will take into account the possible antagonistic, additive, or synergistic effects which may occur between the various toxic compounds present in the mixture of chemicals present in the drinking water supplies under study.

REFERENCES

Craun, G.F. (1981), "Outbreaks of Waterborne Disease in the United States: 1971-1978," J. Am. Water Works Assoc. 73(7):360-369.

Flosom, A.R. and R.J. Prineas (1982), "Drinking Water Composition and Blood Pressure: A Review of the Epidemiology," Am. J. Epidem. 115:818-832.

National Academy of Sciences (1980), Drinking Water and Health, (Washington, DC, National Academy Press; 1980), pp. 5-24.

Neal, R.A. (1982), "Evaluating Potential Health Risks of Consuming Reused Water," J. Am. Water Works Assoc. 73:638-641.

CHAPTER 2

DRINKING WATER: A GLOBAL VICTUAL

J. R. Hickman, B.Pharm, M.S.

Department of National Health and Welfare
Health Protection Branch
Ottawa, K1A 0L2
Canada

There are three requisites for human life: a supply of air, an adequate quantity of nutritious food, and water to drink. But while some societies are nourished mainly by rice, or meat, or fish depending on local tradition and availability, all depend on water, the global victual.

Essential to life, water is one of the earth's most abundant substances. We find it everywhere -- in the sky, on the ground and underneath it as a liquid, a solid and a gas. But you will not find pure water anywhere in nature. Water carries and dissolves so many other substances that is has been termed the universal solvent. In the hydrologic cycle, it absorbs gases such as sulfur dioxide and oxides of nitrogen present in the atmosphere as it returns to the earth's surface where, as water or ice, it erodes and dissolves the rocks and soils and picks up other natural pollutants as wilderness streams to flow to the oceans.

Although mention of the presence of chemicals in drinking water usually leads to thoughts of potential adverse health effects, we should not lose sight of the beneficial qualities that chemical constituents bestow on drinking water.

Anyone who has tasted chemically pure water will agree that it is insipid, tasteless and flat. Good-tasting water depends upon its ionic, inorganic constituents for its aesthetically pleasing qualities. In fact, from a rigorous point of view, it has been suggested that the taste of drinking water can be defined as "the sensation that is due to the presence of substances in water which have negligible vapor pressures and negligible odors" (Anon.,1980).

The presence of small amounts of fluoride in drinking water leads to a definite, substantial reduction in the incidence of dental caries, especially in children. Natural waters contain from trace amounts to 10 mg/L (and occasionally, even higher). Whereas a small amount of fluoride is beneficial, higher concentrations (above 1.5 mg/L fluoride) may lead to mottling of the teeth (dental fluorosis), skeletal fluorosis (3-6 mg/L) or, with very high levels, crippling fluorosis (WHO, 1970). Generally, higher concentrations tend to be associated with underground sources and in some regions (in some parts of India, for example) natural fluoride levels are so high as to present public health problems. Most surface waters, however, contain well below 1 mg/L fluoride and many communities have chosen to practice fluoridation of water supplies as a public health measure.

Another example of the beneficial properties of chemical constituents in water is the negative statistical correlation that has been observed between hardness and cardiovascular disease. The first report of a close association between death rates from strokes and the acidity of river water-derived drinking water came from Japan in 1957 (Marier et al, 1979). Since that time, a number of studies from various parts of the world have demonstrated a statistical association between "soft" water and cardiovascular disease.

In one such undertaking, the British Regional Heart Study, cardiovascular mortality was studied in relation to water quality (hardness and more than 20 other water parameters were studied), climate, air pollution and genetic factors in 253 British towns. After adjustment for other factors, soft water areas (around 0.25 mmol/L) were found to have a 10-15% higher cardiovascular mortality than areas of medium hardness (Pocock et al, 1981). Others have been unable to demonstrate a statistical relationship between hardness and cardiovascular disease (Zeilhuis and Haring, 1981) but, as noted by Zoeteman (1981), countries such as the Netherlands, not having very large areas supplied with very soft water, may simply not be suitable for investigating such statistical relationships.

Several hypotheses have been proposed in attempts to account for the relationship but, at present, there is no definite evidence that calcium and magnesium are involved. The two most quoted hypotheses relate to (1) constituents in hard water being protective in some way, and (2) substances in soft water (e.g., metals leached from piping materials), promoting the disease. In the case of the protective hypothesis, it is often considered that the diet provides an adequate supply of calcium and magnesium, although for magnesium, there is the possiblity of a dietary deficiency in some situations (Neri, 1978).

To most people, however, the mention of chemicals and drinking water conjures up visions of pollution and associated subtle ill-health manifestations. Increasingly, public health authorities are becoming concerned by the possibility of water we drink being the vehicle whereby human beings are exposed to untoward quantities of potentially toxic substances. Of special relevance are recent epidemiological studies noting possible correlations between increased incidences of a number of diseases including bladder, colon and rectal cancer and the consumption of water from particular sources (Crump and Guess, 1980).

The real problem lies with man's growing use of fresh drinkable water for a myriad of other purposes. In Canada, for example, water withdrawals for all purposes amount to a staggering 7,100 liters per person per day, most of which is used for purposes such as cooling, irrigation, as a medium for waste disposal and as process water in industrial processes. Unlike other natural resources, water has been regarded traditionally as a free good: it is rarely subject to a commodity charge, such charges that have been levied usually being associated with pumping, treatment and distribution. Free goods tend to be overused and misused as were the common pasturelands in medieval England. Further, until recently, it was generally believed that waters had an unlimited capacity for self-purification.

The apparent deterioration in quality of many raw water sources reflects, on one hand, the continually rising demands placed on limited available water resources and, on the other hand, the worldwide increase in the use of chemicals in our modern society. Although long-term data on changes in water quality are scarce, it is interesting to note that chloride concentrations in the Rhine increased linearly from about 15 mg/L in 1875 to about 200 mg/L in 1975 (OECD, 1982) and that annual ice layers from areas of Greenland, identifiable back to about 1200 A.D. reflect an increase in metallic contamination of 70% to 1870, of 210% by the 1950s and of 460% by the 1960s (OECD, 1982).

ORIGINS OF CHEMICALS OCCURRING IN DRINKING WATER

Table 1 shows three broad categories of substances affecting the quality of raw water sources: the "naturally-occurring" substances, point sources of pollution and non-point (or diffuse) sources.

"Naturally-occurring" substances include minerals leached from the underlying geological formations, such as calcium and fluoride as well as organic substances such as humic acids contributed from soils and sediments. The interactions of acidic deposition (acid rain) with the terrestrial ecosystem, including vegetation, soil and bedrock, have resulted in alterations in the concentrations of waters in acid-sensitive areas.

Point sources of pollution include industrial effluents, leachates and run-off from mining sites and land disposal sites of domestic and industrial wastes. Although municipalities increasingly treat sewage, some pollutants, mainly the non-biodegradable organic and inorganic

TABLE I. ORIGINS OF CHEMICALS COMMONLY OCCURRING IN DRINKING WATER

A. Substances affecting the source quality (raw water)

"Naturally-occurring" substances

- leached from geological formations (e.g., calcium, heavy metals)
- derived from soils and sediments (e.g., humic substances)

Pollutants derived from point sources

- domestic sewage treatment (e.g., NTA)
- industrial effluents (e.g., synthetic organics, metals, cyanide)
- landfill waste disposal (e.g., metals, synthetic organics)

Pollutants derived from non-point sources

- agricultural run-off (e.g., fertilizers, pesticides)
- urban run-off (e.g., salt, PAHs)
- atmospheric fall-out (e.g., PAHs, chlorinated organics, heavy metals)

B. Substances resulting from treatment

Substances formed during disinfection (e.g., trihalomethanes, chlorophenols)

Treatment chemicals (e.g., chloramines, fluorides) and their impurities (e.g., acrylamide monomer, carbon tetrachloride)

C. Substances arising from the distribution and service systems

Contaminants arising from contact with construction materials and protective coatings (e.g., lead, vinyl chloride monomer and asbestos fibers from piping, cadmium from fittings, PAHs from coal tar linings)

Substances arising from point-of-use devices (e.g., sodium, silver)

materials may pass through the treatment system without being eliminated, unless advanced technology is applied.

Diffuse sources contribute significantly to the "background" composition of drinking water supplies. In some river basins, diffuse sources may contribute up to half of the total pollution load (OECD, 1982). Run-off from urban and industrial areas contributes significantly to the heavy metal content and other contaminants including asbestos, hydrocarbons and PAHs (Science Advisory Board, IJC, 1982). Over one-quarter of the world's zinc consumption (which also often contains significant amounts of cadmium) is used for galvanizing which erodes and is gradually washed off in the weathering process. Salt used extensively in cold countries in order to prevent icing of roads and highways can add significantly to the chloride concentrations in waters at the time of the spring thaw, and acidic air pollutants, mostly generated in urban areas, may hasten corrosion and lead to secondary pollution.

In rural areas, pesticides and fertilizers may impact on surface and groundwater quality. For example, rapidly rising nitrate levels in aquifers in parts of New Zealand have been linked to increased applications of nitrogen fertilizers and irrigation of horticultural and crop lands (Burden, 1983). Similar problems have been observed in many other countries (Fraser and Chilvers, 1981) and infantile methaemoglobinaemia, resulting in death in about 8% of affected children, has been associated with areas in which high nitrate concentrations are present in the drinking water.

TREATMENT

Because of the recognized problems of contamination of water sources, most surface water sources and some groundwater sources receive some form of treatment prior to their distribution to consumers. Traditionally, such treatment has been applied primarily to reduce the possibility of spreading water-borne pathogenic bacteria and to reduce the content of suspended materials.

In the past decade, a major area of health concern has been the realization that compounds generated in drinking water during treatment may present health hazards.

Chlorine is used widely as a disinfectant and as an oxidant to degrade organic substances and ammonia. It is cheap and convenient to use either as chlorine gas or liquid, or as hypochlorite. In 1974, Rook published his classic paper in which he showed trihalomethanes at significant levels following chlorination at the Rotterdam water utility (Rook, 1974).

Subsequently, the National Cancer Institute announced in 1976 that chloroform (the predominant THM) in most water supplies) is carcinogenic in both rats and mice (NCI, 1976). Gradually, in the intervening period, it has been realized that THMs are only part of the chlorinated byproducts, many of those identified to date being compounds known, under certain conditions, to be toxic.

Alternatives to chlorination most frequently considered are the use of ozone and chlorine dioxide. Ozone has been employed extensively in some countries, notably France, for almost eighty years. It is a more powerful oxidizing and disinfecting agent than chlorine but, unlike chlorine, has no residual disinfectant action in the distribution system. On-site generation equipment is required and the overall process is more expensive than chlorination. Ozonation reaction products are still relatively unknown (Fiessinger, et al, 1981); epoxy by-products have been postulated and extracts prepared from supplies treated with ozone have been shown to be mutagenic (Nestmann, 1983). Pure chlorine dioxide does not give rise to THMs, but a major disadvantage is the production of chlorate and chlorite which may be of health significance when ingested regularly (Robeck, 1981).

In modern treatment technology, a wide range of chemicals may be used, including salts of iron and aluminum, flocculation adjuncts, activated silicates, alginates, synthetic polyelectrolytes, pH adjusting materials and disinfectants. Many of these, particularly the so-called "technical" grades, may be sources of contamination. The EPA is developing a Water Chemicals Codex to specify the purity of many of these chemicals (Rehwoldt, 1982). Another approach endorsed by OECD (1982) requires that no more than 10 per cent of the maximum contaminant levels specified in national drinking water standards should be contributed by treatment chemicals.

DISTRIBUTION

The remaining source of chemical contaminants in drinking water is the distribution network. In the course of its treatment and distribution, water comes into contact with a variety of structures including storage tanks, piping and plumbing fixtures. As a consequence, quantities of adventitious materials may be present in the water (including asbestos, metals, and organics).

Asbestos-cement piping may contribute to the asbestos content of drinking water, particularly when the water supply has aggressive qualities. In Winnipeg, for example (a city with aggressive water), some samples from the distribution system contained 6.5×10^6 fibers/liter at the water works (Health and Welfare Canada, 1979). It is estimated that there are more than 1.5 million miles of A-C pipe world-wide, of which, 200,000 miles are in the United States (Olson, 1974). Although several epidemiologic studies have been reported, all but one were ecological in nature and insensitive because of a large number of confounding variables (Toft et al, 1983). In one case-control authors noted various inconsistencies in their results but concluded that their study did not provide evidence of a cancer risk due to the ingestion of drinking water (Polissar et al, 1982).

Metals. Uptake of metals from pipes, storage tanks and plumbing fittings is a universal occurrence. Concentrations of metals such as copper, cadmium, zinc or lead usually are low, but can become elevated when water stands for a long time in the pipes, especially in new pipework lacking protective deposits. Soft, acidic waters distributed in older systems containing lead-lined storage tanks and lead pipes have resulted in concentrations of lead substantially greater than 100 microg/L in the United Kingdom (Mathew, 1981) and elevated levels of several metals have been found, after standing, in cottages drawing water from lakes affected by acid rain (Méranger, 1983).

Organics. Plastic structures also can contribute to contamination of drinking water supplies through leaching. Examples include vinyl chloride from certain grades of PVC piping, although PVC piping that meets standards and specifications such as those of the National Sanitation Foundation should not give rise to this kind of problem.

SIGNIFICANCE OF CHEMICALS IN RELATION TO HEALTH

In concluding, it is pertinent to consider the impact of chemicals in drinking water with regard to human health. As a result of advances in analytical methodology during the past twenty years, large numbers of organic compounds have been identified in drinking water at relatively low concentrations (microg/L range). In 1979, scientists from fourteen countries, under the auspices of the NATO/CCMS Drinking Water Pilot Study, pooled current knowledge which resulted in a list of 744 chemical compounds identified in drinking water (Borzelleca, 1981). Kraybill (1982) reported that 765 chemicals have been identified in drinking water, of which 20 are recognized carcinogens, 23 suspect carcinogens, 18 are carcinogenic promoters and 56 are mutagenic compounds. The majority of compounds identified to date are volatile, non-polar compounds; however, these represent less than 10% of the dissolved organic carbon (DOC) content. Present attention is shifting towards the non-volatile compounds which make up 80-90% of the DOC, using techniques such as HPLC.

With estimates of known organic compounds now around 1,000 (Kool et al, 1982), the task of assessing the potential for human toxicity on a chemical-by-chemical basis is becoming unmanageable. Neal (1983) recently has published proposals whereby limited toxicological and analytical resources would be directed initially to those organic chemicals that potentially pose the greatest risk to humans based upon lipid solubility, molecular weight and occurrence in water at concentrations exceeding certain defined levels. Neal (1983) proposes that the toxicity of compounds identified by this approach could be evaluated individually or by administration as an artificial mixture. One long-term study using mixtures of chlorinated hydrocarbons frequently detected in drinking water has been carried out by Van der Heijden and van Esch in the Netherlands and briefly reported upon by Kool (1982); no increases in tumor incidence were observed.

In a variation of this technique, Chu et al (1981) administered inorganic substances in combination at levels up to 25 times the IJC objective levels to rats for 13 weeks with no observable effects. More recently, similar investigations using organic substances in mixtures at up to one thousand times objective levels have been fed without adverse effects (Villeneuve, 1983). The absence of any observable toxicity at these levels is a clear demonstration of their suitability to protect health and provides an answer to those who criticize the objectives on the grounds that deleterious interactions may occur.

Various attempts to obtain organic concentrates and use these in toxicological studies have been reported from the U.S., France, Japan, and the Netherlands. These have been extensively reviewed by Kool et al (1982) and have involved testing concentrates prepared by chloroform and ethanol extraction, activated carbon adsorption, reverse osmosis and XAD resins. In severals of these studies, significant increases were observed in the incidence of tumors among animals receiving the concentrates, although several of the studies could be faulted due, for example, to lack of adequate controls.

Scientists in many countries have used mutagenicity tests as a means of screening drinking water and extracts and concentrates prepared from drinking water, in order to identify components showing biological activity. Such studies provide insight and a means of setting priorities in the difficult analytical task of identifying organic compounds of interest.

The difficulties associated with the use of mutagenicity tests in this manner have been pointed out (Harrington et al, 1983; Nestmann, 1983) and generalizations are difficult, if not impossible, because of differences in concentration techniques and the choice of bacterial strains for testing. Loper (1980) pointed out that most known mutagens would escape detection if water containing levels of 1 micorg/L or less were tested directly. However, concentration to the extent of up to 5 x 10^4-fold often is practiced (Nestmann, 1983).

A number of reports suggest that chlorination usually leads to an increase in mutagencity (Loper, 1980; Nestmann, 1983), thus supporting the need to study alternate methods of disinfection. A few Canadian water treatment plants disinfect with ozone and it is of interest to note that extracts of these supplies show mutagenic response (Nestmann et al, 1983), although it must be pointed out that post-treatment chlorination is used to maintain a chlorine residual in the distribution system. Burleson and Chambers (1982) showed that ozone treatment could inactivate the mutagenicity of polycyclic hydrocarbons and aromatic amines, but that other mutagenic compounds were not affected.

Kool et al (1982) list 45 organic mutagens that have been identified in drinking water, but little has been reported to establish correlations between observed mutagenic responses in concentrates of drinking water in terms of organics present. Kool et al (1981) noted that mutagenic activity in drinking water in the Netherlands is mainly in the non-volatile fraction and that these compounds are different from those already identified by GC/MS techniques.

Nestmann et al (1983) recently have published the results of a survey involving water samples from 29 municipalities (30 treatment plants) throughout Canada. At each location, samples of raw water and finished drinking water were compared and each site was surveyed in both summer and winter. The mutagenicity testing of extracts was accompanied by chemical analysis of volatile organics and of triaryl/alkyl phosphates. The highest correlations were found for nonhalogenated organics detected in the XAD-2 effluent, with correlation coefficients greater than 0.9 for benzene, ethylbenzene, toluene and the xylenes. The highest correlations for chlorinated compounds occurred with 1,2-dichloroethane, tetrachloroethylene and dichloromethane; much lower correlation values were found for THMs and 1,2-dichloroethylene. Mutagenicity did not correlate well with total (dissolved) organic carbon.

Finally, a word about epidemiological studies. Several studies have suggested a possible correlation between organics in drinking water and elevated cancer mortality rates (Kool et al, 1982; Crump and Guess, 1980; Council on Environmental Quality, 1981). However, all studies to date have been limited by design flaws (Wigle, 1983). Nevertheless, such studies provide important insights, especially in unravelling the potential cancer risks associated with chemicals in drinking water. When taken in conjunction with the known occurrence of mutagenic substances in drinking water concentrates and the identification of known and suspect carcinogens, such observations from epidemiological studies lend weight to the need to develop new standards for toxic chemicals, for new public policies relating to the safeguarding of our highest quality water sources and for requiring research and development of alternative treatment technologies that result in the highest possible quality of drinking water for all at an affordable price.

REFERENCES

Anonymous (1980), "Guidelines for Canadian Drinking Water Quality: Supporting Documentation," (Ottawa, Canada, Health and Welfare Canada),p. 52.

Borzelleca, J.F. (1981), "Report of the NATO/-CCMS Drinking Water Pilot Study on Health Aspects od Drinking Water Contaminants," Science of Total Environment. 18:205-217.

Burden, R.J. (1983), "Nitrate Contamination of New Zealand Aquifers," Water Quality Bull. 8:22-25.

Burleson, G.R. and Chambers, T.M. (1982), "Effect of Ozonization on the Mutagenicity of Carcinogens in Aqueous Solution," Environ. Mutagenesis 4:469-476.

Chu, I., Villeneuve, D.C., Becking, G.C. and Lough, R. (1981), "Subchronic Study of a Mixture of Inorganic Substances Present in the Great Lakes Ecosystem in Male and Female Rats," Bull. Environ. Contam. Toxicol. 26:42-45.

Council on Environmental Quality (1981), "Contamination of Ground Water by Toxic Organic Chemicals," (Washington, D.C., U.S. Government Printing Office).

Crump, K.S. and Guess, H.A. (1980), "Drinking Water and Cancer: Review of Recent Findings and Assessment of Risks," Contract report for U.S. Council on Environmental Quality (Contract No. EQ10AC018), Washington, D.C.

Fiessinger, F., Richard, Y., Montiel, A. and Musquère, P. (1981), "Advantages and Disadvantages of Chemical Oxidation and Disinfection by Ozone and Chlorine Dioxide. Sci. Total Environ. 18:245-261.

Fraser, P. and Chilvers, C. (1981), "Health Aspects of Nitrate in Drinking Water," Sci. Total Environ. 18:103-116.

Harrington, T.R., Nestmann, E.R. and Kowbel, D.J. (1983), "Suitability of Modified Fluctuation Assay for Evaluating the Mutagenicity of Unconcentrated Drinking Water," Mutat. Res. 120:97-103.

Health and Welfare Canada (1979), "A National Survey for Asbestos Fibers in Canadian Drinking Water Supplies," Report 79-EHD-34, (Ottawa, Canada, Environmental Health Directorate, Health and Welfare Canada).

Kool, H.J., Van Kreijl, C.F. and Zoeteman, B.C.J. (1982), "Toxicology Assessment of Organic Compounds in Drinking Water," CRC Critical Reviews in Environ. Control. 12(4):307-357.

Kool, H.F., Van Kreijl, C.F., Van Kranen, H.J. and de Greef, E. (1981), "Toxicity Assessment of Organic Compounds in Drinking Water in the Netherlands," Sci. Total Environ. 18:135-153.

Kraybill, H.F. (1982), "Carcinogenesis of Synthetic Organic Chemicals in Drinking Water," J. Am. Water Works Assoc. 73:370-372.

Loper, J.C. (1980), "Mutagenic Effects of Organic Compounds in Drinking Water," Mutation Res. 76:241-268.

Marier, J.R., Neri, L. C. and Anderson, T.W. (1979), "Water Hardness, Human Health and the Importance of Magnesium," Report No. 17581 (Ottawa, Canada, National Research Council National Research Council Canada).

Mathew, G.K. (1981), "Lead in Drinking Water and Health," Sci. Total Environ. 18:61-75.

Méranger, J.C., Khan, T.R., Vairo, C., Jackson, R. and Li, W.C. (1983), "Lake Acidity and the Quality of Pumped Cottage Water in Selected Areas Of Northern Ontario," Intl J. Environ. Anal. Chem. 15:185-212.

NCI (1976), "Report on the Carcinogenesis Bioassay of Chloroform," (Washington, D.C., National Cancer Institute, March 1, 1976).

Neal, R.A. (1983), "Drinking Water Contamination: Priorities for Analysis of Organics," Environ. Sci. Technol., 17:113A.

Neri, L.C. and Johansen, H. (1978), "Water Hardness and Cardiovascular Mortality," Ann. N.Y. Acad. Sci., 304:203-219.

Nestmann, E.R. (1983), "Mutagenic Activity in Drinking Water", in Carcinogens and Mutagens in the Environment, Vol. III, Naturally Occurring Compounds: Epidemiology and Distribution, H.F. Stich, Ed. (Boca Raton, FL, CRC Press, Inc.)

Nestmann, E.R., Otson, R., LeBel, G.L. and Williams, D.T. (1983), "Correlation of Water Quality Parameters with Mutagenicity of Drinking Water Extracts," in Water Chlorination: Environmental Impact and Health Effects, Vol. 4, R.L. Jolley, W.A. Brungs, J.A. Cotruvo, R.B. Cumming,

J.S. Mattice, and V.A. Jacobs, Eds. (Stoneham, MA, Ann Arbor Science Publishers, Inc.), pp. 1151-1163).

OECD (1982), Control Policies for Specific Water Pollutants. (Paris, France, Organization for Economic Cooperation and Development).

Olson, H.L. (1974), "Asbestos in Potable Water Supplies," J. Am. Water Works Assoc. 66:515.

Pocock, S.J., Shaper, A.G. and Packham, R.F. (1981), "Studies of Water Quality and Cardiovascular Disease in the United Kingdom," Sci. Total Environ. 18:25-34.

Polissar, L., Severson, R.K. Boatman, E.S. and Thomas, D.B. (1982), "Cancer Incidence in Relation to Asbestos in Drinking Water in the Puget Sound Region," Amer. J. Epidemiol., 116:314.

Rehwoldt, R. (1982), "Water Chemicals Codex," Environ. Sci. Technol. 16:616A-618A.

Robeck, G.G. (1981), "Chlorine, Is there a Better Alternative?", Sci. Total Environ. 18:235-243.

Rook, J.J. (1974), "Formation of Haloforms During Chlorination of Natural Waters," Water Treat. Exam. 23:234-243.

Science Advisory Board, I.J.C. (1982), Annual Review: Great Lakes Reseach (Ottawa, Canada and Washington, D.C., International Joint Commission).

Toft, P. and Meek, M.E. (1983), "Asbestos in Drinking Water: A Canadian View," Environ. Health Perspectives 53:489-498.

Villeneuve, D.C. (1983), Personal Communication.

WHO (1970), Fluorides and Human Health. Monograph Series No. 59. (Geneva, Switzerland, World Health Organization).

Wigle, D.T. (1983), "Drinking Water and Human Cancer: A Brief Review of Epidemiologic Studies," Report to IARC, Lyon, March 1983.

Zeilhuis, R.L. and Haring, B.J.A. (1981), "Water Hardness and Mortality in the Netherlands," Sci. Total Environ. 18:35-45.

Zoeteman, B.C.J. (1981), "Water Supply and Health," Sci. Total Environ. 18:363-366.

CHAPTER 3

PUBLIC DRINKING WATER AND CHEMICALS

James F. Manwaring, P.E.

American Water Works Association Research Foundation
6666 West Quincy Avenue
Denver, Colorado 80235

The condition of the nation's water supply is a topic of keen interest to those in the business of providing drinking water to the public. This conference represents a wide spectrum of interests and responsibilities and it is salutary indeed for us to get together and share our perspectives. My part in the program is to provide the viewpoint of the tens of thousands of water supply professionals, utilities and other organizations who comprise the membership of the AWWA Research Foundation.

The AWWA Research Foundation is a nonprofit organization whose purpose is to manage a coordinated national research and development program for the water supply industry. The Foundation selects and funds research projects designed to develop or improve technologies that will help the water supply industry meet present and future drinking water needs in environmentally and economically acceptable ways. The Research Foundation's activities are coordinated with those of government agencies, individual utilities, manufacturers, universities, and comparable organizations in many other countries.

The theme of this conference, "The Impact of Chemicals on a Limited Resource," is very interesting when taken from the perspective of the water supplier. The impact of chemicals may be either beneficial or detrimental or both, depending upon one's point of view. On one hand, chemicals make it possible to treat water from other than pristine sources to make it palatable and nonharmful to the consumer; in this sense chemicals have expanded the usability of a limited resource. On the other hand, chemical contamination of some water sources has eliminated or minimized their benefical use. Obviously, the chemical sword has two edges.

The perspective of the water supplier is that he must take his water as it comes, good or bad, treat it to the best of his ability, and deliver a product that satisfies both consumer tastes and standards of public health protection. Naturally he is interested in quality of the source and its protection, but by and large he has little power beyond the boundaries of his own storage and operations.

BENEFICIAL IMPACTS

In the U.S. an estimated 1 to 2 billion gallons of drinking water is provided each day to the customers of public water systems. To comply with health and other standards for treating that amount of water, it is estimated that over $272 million was spent on the purchase of chemicals by the U.S. water utilities in 1982 (1). Table I shows the quantities of chemicals that were used by the water industry in 1982.

The most beneficial chemical in the history of water supply has been and will continue to be chlorine. Chlorine is economical, effective, convenient, and its proper use has virtually eliminated the transmission of bacteriological and viral diseases by drinking water. Despite chlorine's assocition with THM production, its benefits far outweigh any potential adverse characteristics. Chlorine is virtually the only irreplaceable chemical in the water purveyor's arsenal of weapons against disease transmission via drinking water. The application of chlorine has vastly expanded the usability of our water resources by allowing the use of other than pristine and protected sources.

TABLE I. CHEMICALS IN WATER TREATMENT (1982 - 1000 tons) (1)

Coagulation		Softening	
Alum	250	Calcium oxide	350
Ferric chloride	80	Hydrated lime	170
Ferric sulfate	100	Sodium hydroxide	135
Polyelectrolytes	4.5	Carbon dioxide	18
		Soda ash	200
Disinfection		**Miscellaneous**	
Chlorine	500	Fluoride compounds	38
Hypochlorite	2.5	Activated carbon	10
Sodium chlorite	6.4	Phosphate	16
Ammonia	2.5	Sodium chloride	6.5
		Copper sulfate	1.5

No less significant is the use of coagulants within the water industry. Essentially there are no surface water sources which could be used without the application of some type of coagulant to remove turbidity and some inorganic contaminants. These are other examples of chemicals which expand the usability of our water resources. The softening chemicals certainly have acted in the same beneficial way by modifying the quality of the water so that it can be used for more purposes.

While fluoridation cannot be pointed out as increasing the resource availability, the benefits of fluoridation in terms of reducing tooth decay cannot be disputed. This provides just another aspect to the benefits of chemicals on a limited resource.

The water industry is just now in the beginning phases of investigating wastewater reuse as a viable resource alternative for future drinking water supplies. Already water reuse has been shown to be effective and economical for irrigation, industrial, and other uses lower than human consumption. In all these cases chemicals play an important role in the reuse train. Reuse is nothing more than increasing the utilization of a limited resource.

DETRIMENTAL IMPACTS

Water supply systems today are beset by an array of contamination problems which, during the past decade, have been growing in volume and complexity. Hundreds, if not thousands, of new chemical compounds are developed every year. Many of them find their way into our water. We know little or nothing about their potential health effects and we know less than that about their effects in combination with each other. No local research program in the world has the resources to monitor for all the compounds found in drinking water, to examine them in combination, to perform toxicological tests on all of them, and to conduct epidemiological studies on all of them. It is no exaggeration to say that the dimensions of the task boggle the mind as well as the pocketbook.

This is why the water industry looks to the government for health effects research. Basically, we are engineers, not toxicologists or epidemiologists. Tell us that something must be removed from the water, and we'll find a way to control it and tell you what it will cost. But in the same breath with which we call for more health effects research, we emphasize thorough and careful research, the results of which should be made fully public and be able to withstand the scrutiny of scientific review. Like other citizens, the water supplier demands the scrupulous caution of the true scientist. On the one hand, if there is substantial evidence that a constituent in drinking water constitutes a threat to human health, we want to remove it. On the other hand, we want to be certain of the facts, for the cost of monitoring and controlling a large number of chemicals measured in parts per billion would impose an extraordinary burden on the consumer -- a burden that might well be disproportionate to the protection gained.

The point is that the regulation of contaminants in drinking water must be a public decision based solidly on data that are made public, with the public involved in reviewing the risks and costs involved.

WATER SOURCE CONTAMINATION

The problem of, and solution to, inorganic chemical contaminants has received a great deal of attention in the past and will not be repeated here; the increased number of organic contaminants being detected in both surface and groundwaters throughout the world will be the focus of the remainder of this paper. The organic contaminants receiving the most attention currently are referred to as volatile organic chemicals (VOCs). They are named VOCs because of their distinctive common property of high volatility relative to organic substances such as pesticides.

Most water supply personnel are familiar with man-made organic substances, such as phenols and chlorophenols, because of the adverse taste-and-odor problems they cause. Also, regulated pesticides are a family of man-made organic substances familiar to waterworks operators. The presence of these substances, although undesirable, does not result in questions regarding effective types of removal, or magnitude of the costs that may be incurred, since experience has provided acceptable answers to these questions.

Generally, contaminated surface waters have been found to contain a broad spectrum of organic substances of very low concentration (2), often in the nanogram per liter range. Conversely, contaminated groundwaters may contain a lesser number of compounds that occur at very high concentrations (as high as 400 mg/L) (2-6). Data from organic contamination surveys by state agencies and the U.S. Environmental Protection Agency (USEPA) have begun to change the general perception of groundwater quality. The long accepted viewpoint that groundwater sources are of pristine purity and in need of minimal treatment is being questioned in current water quality forums (7-10).

USEPA has conducted three major water supply surveys that have generated a great amount of data on organic contamination of drinking water supplies (2-4). Each survey was designed to focus upon selected contaminants, geographic distributions, and water utility sizes, and to provide a broad range of information. None of the surveys focused upon specific locations for detailed temporal data. A review of these data for surface water supplies was provided in a 1981 EPA publication (5). The data indicate that trichloroethylene and carbon tetrachloride appeared in about one-third of all samples analyzed. Typically, mean values of samples containing VOCs (positive samples) ranged from

0.5 ppb to 3.5 ppb; maximum values fell in the range of 10 times the average values.

Data from the EPA surveys provide an adequate overview of the national situation in the U.S., but do not address the major threat posed to surface streams by VOCs: that is, infrequent, short-term occurrences of high concentrations caused by spills and deliberate discharges. This situation can be documented only by collecting samples at a frequency that is comparable to the expected duration of the event. The most publicized spill in the U.S. occurred in February 1977 on the Kanawha River, a tributary of the Ohio River located in West Virginia. An estimated 140,000-lb discharge of carbon tetrachloride was observed during what was originally a routine analysis in the NOMS program (11). Carbon tetrachloride was detected at 50 ppb in a single grab sample at Huntington, West Virginia on the Ohio River. A lack of data made the determination of maximum concentrations difficult. The spill incident did result in the improvement of monitoring and communication networks in the Ohio River Valley.

Contamination of groundwater by organic chemicals is much more commonplace than previously believed. With groundwater, relatively small sources of pollutants can cause a high degree of contamination to a localized area without affecting surrounding areas. Continuous discharge may cause mounding and shift the direction and speed of water flow somewhat unpredictably.

The lack of predictability of contamination in aquifers and the great reliance of small communities on groundwater supplies combine to make an overall assessment of groundwater contamination difficult. Early efforts in groundwater monitoring were focused in areas of heavy industrialization where the use and disposal of industrial chemicals occurred in high volume.

Surveys conducted by individual states have yielded a great deal of data on specific sites of contamination. The only national overview of the problem in the U.S., however, has been provided by the recently completed Ground Water Supply Survey (GWSS) conducted by EPA in conjunction with the states (12). The data from the GWSS was divided into two subclasses: data from randomly selected sample locations, and data from locations thought to have a high probability of contamination, but for which no data were yet available.

The summary data indicated that 16 to 37% of all samples contained at least one of the measured VOCs, depending upon the subclass. Detections occurred at a significantly higher frequency and concentration in the non-random samples as opposed to the random samples, as would be expected. These data reflect a great deal of similarity to previously reported data for surface streams, with the exception that the maximum values encountered are 50 to 100 times the average values of positive detection.

It is stressed that the GWSS included sample locations that were either random or that were selected on the suspicion of contamination without existing data. These qualifications largely excluded groundwater supplies that were known to be contaminated to some extent by VOCs. An appreciable amount of data has been collected by state surveys in areas contaminated with VOCs. Although statistical interpretation of these data is difficult because of varying conditions of sample collection and analysis, the data do present valuable information on seriously contaminated aquifers within limited geographic confines.

The problems associated with groundwater contamination constitute a worrisome litany:

- Contaminant concentrations, particularly for the organic chemicals, tend to be at much higher levels in groundwaters than in surface waters.

- Contamination is difficult and costly to detect.

- Contaminant removal can be so expensive and technically complex that it is often not a viable option, especially for small systems.

- Because of the slow rate of groundwater movement, natural flushing of an aquifer may take hundreds of years; and the source of pollution may be impossible to identify if the problem doesn't surface, so to speak, for decades after it was caused.

- Lastly, the sources of groundwater contamination are as diverse and difficult to control as those affecting surface waters. They include industrial waste disposal and impoundment sites, agricultural practices, some injection practices, depletion and salt water intrusion, urban run-off, interaction with polluted streams, and the cumulative effects of many small, non-point source activities.

Thus far, it seems that pollution has affected only a small fraction of our groundwater resources. What should trouble us, however, is not the degree of contamination, but the fact that it can and does occur and that the conditions which caused the known instances exist elsewhere in the country.

SMALL WATER SYSTEMS

It is noteworthy that the preponderance of small water systems draw upon groundwater sources and that these are the very systems which are patently ill-equipped to cope with contamination. Using round figures, 10% of the nation's water systems serve 90% of the population; 90% of the systems serve the remaining 10% of the population. To be a little more precise, 88.4% of our 59,000 systems serve fewer than 3,300 people. Almost 38,500 systems serve fewer than 500 people. Typically, these are groundwater systems with a pump, a simple distribution system, perhaps a chlorinator, and an untrained part-time operator.

The principal groundwater problems involve nitrates, the naturally occurring contaminants of arsenic, barium, fluoride, radium, and selenium, and a small number of organic pollutants consisting primarily of industrial solvents. These are problems which are minor in nature and easily corrected by the introduction of large-scale treatment facilities; however, this is not practical in many small system situations for three major reasons:

Economics

The cost of installation and operation of the necessary treatment processes normally is beyond the reach of consumers in the small systems. According to a 1980 EPA study, the installation of a treatment plant in a small system would increase the average customer bill from $5.20 per month to $51.00 or more per month, a ten-fold increase. Comparatively, for a large system serving a population of 100,000 people, the same problem and solution would raise monthly bills from $4.20 to $5.60 -- a tolerable cost in the interest of public health protection. The same study concluded that even with a 100% construction grant program (a likelihood in no one's book), the majority of small systems would be unable to meet the standards because they couldn't afford to operate the plant and to pay a competent operator.

Lack Of Appropriate Treatment Technologies For Small Water Systems

Removal technologies certainly are available for all the contaminants of concern, but in many cases those technologies have not been scaled down for the smaller systems. EPA has invested a great deal of research funds in this area over the past few years, but a great deal of work still needs to be done.

Lack Of Well Trained Operators Available To The Small System.

In short, small systems cannot afford to retain operators that are well qualified to operate even a simple chlorinator. AWWA, EPA, and the Rural Water Association all have training and education programs aimed at this class of operator which will, in the long-term, at least reduce, if not eliminate, the situation.

The root cause of the small system problem is related directly to costs. Small systems' revenues are low because of the low number of customers. Consequently, they have to keep rates down as far as possible because often the people they serve are low income or fixed income families. They do not have access to capital for improvements because they have such low revenue and limited assets of any value. Operationally, their low budgets preclude hiring and retaining skilled personnel to staff the system. Often small systems can't afford to replace old equipment or even to maintain the system properly. This leads not only to service breakdowns but also to water quality problems. For certain contamination problems, appropriate technology simply is not available on the small scale required, even if the system could afford it.

EPA and the states are fully aware of these realities, but neither level of government is in a position to do much about the basic causes. This is not to say that regulatory agencies are doing nothing to help small systems. Quite the contrary, many state programs in particular have had a demonstrably positive impact on small system performance. Only the more severe cases fall beyond their influence.

RESPONSIBILITY

All that has been discussed in this paper, by way of representing the viewpoint of the water supplier, serves as background to this fundamental question: who should bear the responsibility for contaminant removal from a drinking water source? The protection of the public health and our water resources is a national goal, and therefore the means of achieving it is a decision that must be made clearly and fairly. It seems to me the following factors should be considered when we try to fix the responsibility, whether it be contamination of groundwater or surface water:

Equity

Should the downstream user bear the cost of removing an upstream discharge? Should discharge permits be based upon different considerations and risks than drinking water standards and health advisories? In many cases the former are based upon best available treatment while the latter are based upon extrapolated health risks. Should the citizens of one community pay the cost of cleaning up the pollution caused by other communities or industries upstream?

Enforceability

Which approach is more enforceable: contaminant removal before discharge or after withdrawal for public consumption? Can government force a community of a few hundred people to undertake water treatment costs that are demonstrably beyond the people's ability to pay? Do you put the town council in jail or shut down the water system, or is it more practical to enforce discharge limitations and standards?

Health and Environmental Protection

At which point - discharge or water treatment plant -- are public health and the environment best protected, not only for drinking water purposes but for other stream and aquifer uses as well? Those concerned about stream ecology or recreation very likely would join in arguing for protection of the source itself, don't you think?

Economics

Which alternative is most cost-effective: treating relatively high concentrations of known contaminants in discharges or monitoring and treating at the water plant for a wide range of unknown contaminants at very low levels? Should the cost of removing wastes be associated with the products being produced or be passed along to those who drink water?

Method of Protection

From the viewpoint of both technology and public policy, is it wise to reply on the water treatment plant to provide the principal, if not total, defense against human exposure to pollutants? The water system is the final point of protection against consumption of harmful chemicals.

Quite obviously, all these factors and issues have broad implications, and they can hardly be resolved unilaterally by the water supplier. Governments at all levels are involved; so are industry, agriculture, recreation, and the general public. It is also important to point out that several federal and state laws and regulatory programs are involved in this effort and should be applied in a coordinated effort. The Safe Drinking Water Act does not stand alone; it is interrelated with the Clean Water Act (discharge permits), Resource Conservation and Recovery Act (groundwater), Toxic Substances Control Act and the Superfund Act.

SUMMARY

This paper has attempted to summarize, from the water utility point of view, the impacts of chemicals on drinking water. In one sense, chemicals are a necessity and have vastly increased the utilization of a limited, yet renewable, resource. In another sense, chemical contamination of drinking water sources certainly has reduced the usability of some water resources for potable purposes. This is particularly true for small water supply systems that do not have the financial nor the technological basis to cope with the problems presented by the presence of the chemicals in the source water.

The water utility manager also must evaluate the question of responsibility with repect to man-made chemical contamination of a drinking water source. The major issue is not one of placing blame on the guilty party, but rather of assigning responsiblity for the protection of the resources and assuring that all the users of that resource are treated in an equitable manner.

What do water suppliers want? The more thoughtful among them are clamoring for more and better research into the health effects of contaminants as a more reliable basis for regulating any additional standards. They ask for cooperation and action from government, industry, agriculture and the public in protecting drinking water sources from pollution and for recognition of the fact that preventive measures may cost less in the long run than the price to be paid for health services, new treatment plants and other effects of contaminated drinking water. Water suppliers also seek a greater and more realistic appreciation of the costs associated with the delivery of a safe, reliable supply of good quality water. Finally, the industry asks that we not lose sight of the fact that among the competing uses for our water resources -- sports and recreation, industry and agriculture -- drinking water must continue to be given top priority. Public health and safety depend upon that commitment.

American drinking water is still, all things considered, the finest quality water in the world. That fact, however, cannot be taken for granted. On the frontiers of the new knowledge science is discovering, there are serious issues requiring national attention, study, and public investment. The science of the last decade has brought us to the threshold of a new era in which we shall learn much more than we know now and in which we shall have to do more than we do now to preserve our proud claim to the best and safest drinking water.

REFERENCES

1. Miller, G.W., "Internal Study for the AWWA Research Foundation", December 1982.

2. Symons, J.M., "National Organics Reconnaissance Survey for Halogenated Organics in Drinking Water." J. Am. Water Works Assoc., 67:(11) (Part I): 634 (1975).

3. National Organics Monitoring Survey (NOMS). Technical Support Division, Office of Drinking Water, U.S. EPA (1978).

4. Community Water Supply Survey (CWSS). Office of Drinking Water, U.S. EPA (Mar. 1981).

5. The Occurence of Volatile Synthetic Organic Chemicals in Drinking Water. Office of Drinking Water, U.S. EPA (Dec. 1981).

6. The Ground Water Supply Survey - Summary of Volatile Organic Contaminant Occurrence Data. Technical Support Division, Office of Drinking Water, U.S. EPA (Jun. 1982).

7. Sayre, I.M., "Organic Contaminants in Groundwater," J. Am. Water Works Assoc., 74(8):381 (1982).

8. Morrison, A. "If Your City's Well Water Has Chemical Pollutants, Then What?" ASCE, 51(9):65 (1981).

9. Volatile Organic Chemicals (VOC) Workshops Conducted by AWWA Research Foundation, under grant from U.S. EPA (1982).

10. Dyksen, J.E. & Hess, A.F., "Alternatives for Controlling Organics in Groundwater Supplies," J. Am. Water Works Assoc., 74(8):395 (1982).

11. Manwaring, J.F., "EPA Puts Emergency Water Provision Into Action," Water and Wastes Engineering, (Apr. 1980) pp. 40-44.

12. Westrick, J.J. "Ground Water Supply Survey, Summary of VOC Occurrence Data," EPA Office of Drinking Water, Cincinnati, OH (June 1982).

CHAPTER 4

BOTTLED WATER: AN ALTERNATIVE SOURCE OF SAFE DRINKING WATER

Jerry T. Hutton, Ph.D.

Vice President, Operations Resources
Government Relations and Scientific Affairs
McKesson Operations Resource Group
One Post Street
San Francisco, California 94104

Americans, in growing numbers, are turning to bottled water to quench their thirst. Last year, the bottled water industry served about 15 billion glassfuls of water to more than 12 million consumers in every state of the Union. According to trade statistics, the number of people drinking bottled water is increasing at more than 10 percent per year. At current growth rates, industry sales could top $1 billion by 1985.

The most frequently asked question about the industry is, why do people buy bottled water? Until recently the answer has been threefold: to satisfy a special dietary need, as a source of safe drinking water in times of emergency, and for aesthetic reasons. Bottled water is recommended and often medically-prescribed for pregnant women and infants in areas where nitrate levels in a local water supply are excessive, for individuals on sodium restricted diets, and to protect children from mottled teeth in regions where fluoride levels are too high.

Use of bottled water in emergency situations caused by floods, earthquakes, and hurricanes is a matter of record. Bottled water also is used as an alternative source of drinking water when the safety of a community water supply is compromised by contaminants of natural or industrial origin. Asbestos in Duluth, radioactivity at Three Mile Island, nitrates and selenium in Wyoming, and pesticides in Hawaii are illustrative of just some of the reasons why and locations where bottled water has proven invaluable in times of emergency -- a use now officially recognized by the Department of Interior and in EPA's National Contingency Plan under the Superfund Act.

Aesthetic factors are the driving force behind most purchases of bottled water. Consumer demand for water that looks clean and tastes good readily explains why bottled water sales are concentrated in geographic areas where the aesthetic quality of water is poor. Per capita consumption of bottled water is greatest in the Southwest, where growing populations and limited water supplies combine to push total dissolved solids (TDS) levels up and consumer acceptance down. For example, bottled water use in Southern California exceeds the national average by sixfold. Nationwide, one consumer in 18 uses bottled water. California, where one of every six consumers drinks bottled water, tops the national average by three-fold. In Southern California, the pace-setter, one of every three consumers uses bottled water both at home and when away from home.

Today, health considerations, in addition to dietary, emergency, and aesthetic factors, are beginning to affect buying habits. Even though the bottled water industry makes no health claims, customer inquiries and consumer interviews reflect a growing public concern about the safety of public and private water supplies.

Although a wide range of package sizes is available, most bottled water users are served by direct delivery of large 5-gallon bottles to homes, offices, factories, hospitals, and public facilities. This means that bottled water tends to be purchased for all members of the family or work-staff, rather than only for those individuals having special dietary needs. This trend is clearly evident in an analysis of the U.S. bottled water sales, which shows that direct delivery of 5-gallon bottles accounts for about 80% of the industy's total estimated 1982 sales of $700 million. Grocery distribution, typically in 1-gallon bottles, accounts for about 20% of total sales.

	U.S. BOTTLED WATER SALES	
	Dollar Sales (Million)	Percent (Approx.)
DIRECT DELIVERY		
Home	420	60
Institutional*	140	20
GROCERY	140	20
Total Sales	700	100

* Offices, Factories, Hospitals, Public Facilities.
Source: 1982 Industry Survey Data.

To support the quality and safety of its products, the industry is organized to ensure that each of its members (a) meets all regulatory requirements, (b) is familiar with new water treatment technology, and (c) employs accepted manufacturing practices.

The bottled water industry is represented by a trade association headquartered in Alexandria, Virginia. Formed in 1958 as the American Bottled Water Association and renamed the International Bottled Water Association in 1981, IBWA represents over 200 bottler members who distribute more than 85% of all bottled water sold in the United States.

The chartered purpose and objectives of the Association are to:

- Work with all legislative bodies and government agencies involved with the supply and regulation of drinking water;
- Conduct continuing education programs to assist industry members with quality control and production procedures;
- Facilitate exchange of technical, scientific, and governmental relations information among members.
- Promote the production and sale of packaged drinking water.

PRODUCTS

The industry distributes four generic types of bottled water, all of which must be derived from a protected source:

- Natural water is water obtained from a protected spring or well.
- Mineral-free water may be produced by distillation or by demineralization.
- Fluoridated water may originate from a natural source or may contain added fluorides in an amount ranging from 0.8 to 1.7 milligrams per liter, depending on the annual average daily air temperature at the location where the bottled water is sold.

PROCESSES

All bottled water is processed in some manner. IBWA membership standards require that any non-chlorinated public water supply be treated by mechanical filtration and ozonation prior to bottling. If a chlorinated water supply is used as the source water, activated carbon filtration and ozonation are required. If the source water is a protected private water supply, such as a spring or well, the minimum treatment required is ozonation or other acceptable means of protecting the product against bacterial contamination.

Beyond these minimum requirements, most bottled water companies employ a more sophisticated range of technologies depending on the composition of the source water and the types of end products produced. Advanced water treatment techniques such as distillation, deionization, and reverse osmosis may be used singly or in combination to produce demineralized water. One measure of the degree of water treatment is found in the fact that the industry removed more than 1.25 million pounds of dissolved chemical substances from its incoming source waters last year. A remineralization process is used to add back low-levels of selected minerals when "good taste" is the sought-for consumer attribute. Ozonation, approved by FDA as GRAS, is used by most bottled water suppliers as a biological control agent.

Packaging is a final safeguard. To assure that its products are fully protected during distribution, the bottled water industry has been and continues to be a pioneer in the design and use of large 5-gallon food containers. At the present time, the industry is actively engaged in converting from glass to a lighter weight, energy conserving and equally protective polycarbonate bottle.

Figure I shows how a bottled water operator might combine several processes in the manufacture of fluoridated water, purified water, and drinking water. If, for example, the source water was chlorinated (e.g., a municipal supply, both mechanical and activated carbon filtration would be used -- the carbon to remove residual chlorine and organics. Thereafter, a softener or other conditioning treatment might be used ahead of reverse osmosis to substantially reduce levels of total dissolved solids. Treatment with a mixed-bed resin is a typical step in the production of purified water from which both fluoridated water and drinking water are manufactured through the addition of appropriate mineral mixes. It is sometimes economically advantageous, depending on the composition of the source water, to use cation and anion exchange resin pretreatment ahead of the mixed-bed resin. In keeping with common industry practice, note that each product is treated with ozone prior to bottling.

REGULATION

Regarded as a food by the Food and Drug Administration, bottled water is the nation's most highly regulated and monitored drinking water supply. To comply with FDA's Quality Standard (21 CFR Section 103.35), and Good Manufacturing Practices (21 CRF Sections 129.1, 129.20, 129.3, 129.35, 129.37, 129.40, 129.80) for bottled water, all industry products must come from protected sources, be bottled in facilities regulated as food plants, processed using manufacturing practices approved by the Federal Government, delivered to consumers in sanitized bottles whose safety is assured by federal regulation, and so labeled as to provide public notification whenever the microbiological, physical, chemical, or radiological quality of any bottled water product is sub-standard.

In addition to FDA, most states now regulate bottled water.

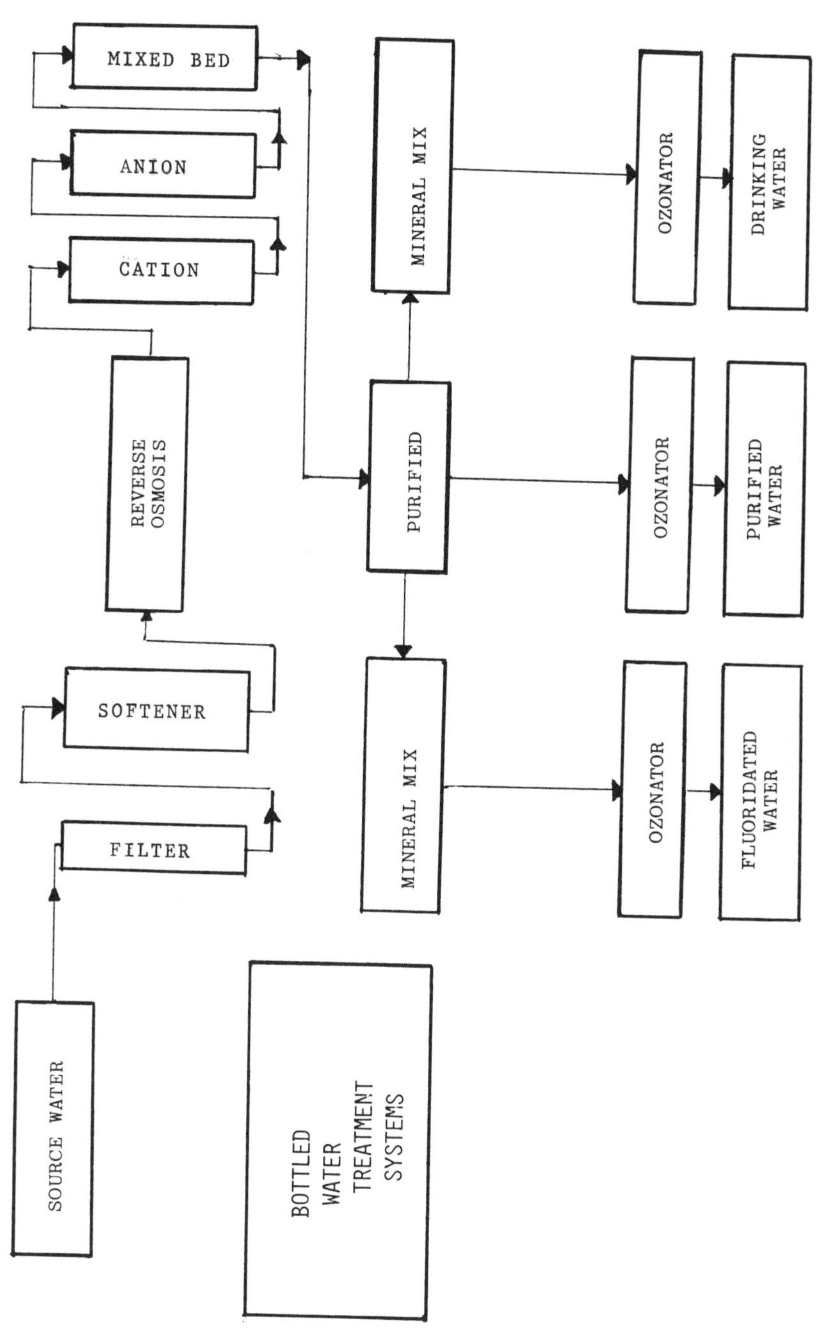

Figure 1. Processes used in producing bottled water.

Under provisions of the Safe Drinking Water Act of 1974, and in keeping with a 1978 Memorandum of Understanding between FDA and EPA, the Environmental Protection Agency also has a definite impact on the regulation of bottled water. The 1974 Act requires FDA to adopt revisions in EPA's National Drinking Water Standards for bottled water, or publish in the Federal Register reasons for not doing so. Taken together with the requirement of FDA's Quality Standard, this FDA-EPA arrangement means that bottled water must meet all currently effective provisions of EPA's Interim Primary and all Secondary Drinking Water Standards (Table I). No other drinking water is so highly regulated. And, needless to say, similar controls and safeguards are neither possible nor economically feasible for all municipal, community, and private water supplies.

Frequency of inspection is a final measure of regulatory oversight. A just completed IBWA survey of member companies shows that bottled water plants throughout the U.S. are inspected at least annually by a federal, state, or local agency. In fact, most plants are inspected several times a year.

SELF-REGULATION

In addition to government regulation, the industry has a comprehensive program of self-regulation. The foundation for this program is the IBWA Plant Technical Manual which provides all members with detailed "how to" information on processing and quality control techniques to be used in the production and bottling of each industry product. To ensure that these procedures are adhered to, the industry employs an independent laboratory -- the American Sanitation Institute -- to audit the performance of each member company. The ASI audit, conducted annually, evaluates compliance with both the Association's performance requirements and FDA's Quality Standard and Good Manufacturing Practices. To stimulate high performance on the part of each member company, the audit results are used as the motivational thrust in IBWA's Product Excellence Award Program. To keep the industry informed on the latest quality control techniques in water processing and bottling, IBWA sponsors an ongoing series of technical seminars and workshops along with an operator certification program. Each of these programs is available to members and non-members alike.

Summarizing to this point, we note that bottled water is a publicly accepted source of drinking water. Each year growing numbers of Americans use bottled water at home and away-from-home locations. While aesthetic factors -- appearance and taste -- are the traditional motivations, health concerns are beginning to influence buying habits.

Because all industry products meet EPA's Interim Primary and Secondary Drinking Water Standards, bottled water is an ideal alternative whenever a public or a private

TABLE I. FEDERAL DRINKING WATER STANDARDS (mg/L)

Substance	FDA Bottled Water Standards	EPA Primary Standards	EPA Secondary Standards
Arsenic	0.05	0.05	
Barium	1.0	1.0	
Cadmium	0.01	0.01	
Chloride	250.0		250.0
Chromium	0.05	0.05	
Copper	1.0		1.0
Iron	0.3		0.3
Lead	0.05	0.05	
Manganese	0.05		0.05
Mercury	0.002*	0.002	
Nitrate (N)	10.0	10.0	
Phenols	0.001		
Selenium	0.01	0.01	
Silver	0.05	0.05	
Sulfate	250.0		250.0
Total Dissolved Solids	500.0		500.0
Zinc	5.0		5.0
Turbidity	5.0	<5.0	
Color	15.0		15.0
Odor	3.0		3.0
Corrosivity	---		Non-corrosive
Foaming Agents	---		0.5
pH	---		6.5 - 8.5
Endrin	0.0002*	0.0002	
Lindane	0.004*	0.004	
Methoxychlor	0.1*	0.1	
Toxaphene	0.005*	0.005	
2,4-D	0.1*	0.1	
2,4,5-TP Silvex	0.01*	0.01	
Radioactivity	Concentration in Picocuries per liter		
Gross Alpha Plus Gross Beta	<5 *	<5	
Trihalomethanes	100 ppb**		

*Effective July 1, 1979
**Federal Register 48(40):68624(November 29, 1979)

water supply fails to meet consumer expectations for health or for aesthetic reasons.

FUTURE ISSUES

Before closing we should ask, what can bottled water contribute to the nation's future safe drinking water programs? As I see them, the major national drinking water issues to be dealt with in the years ahead are safety, availablility, and cost.

Even though progress is being made in the area of safety, much remains to be done. Many questions as to the public health significance of short-term and life-time exposure to the hundreds of organic chemicals known to be present in drinking water are backlogged -- waiting to be answered. Using ever more sensitive detection methods, analytical chemists are discovering organic chemicals in drinking water at rates faster than toxicologists and epidemiologists can assess their health significances. In some regions of the country, acid rain-induced alterations of the natural leaching process represent an uncharted and potential source of toxic pollutants. Finding workable and economically acceptable answers to the problems of heavy metals and other contamination originating within distribution systems will demand a substantial commitment of scientific and financial resources.

The nation's news services serve up steady reminders that drinking water safety is a serious public concern. Media attention to lead, barium, cadmium, TCE, DBCP, EDB, bacteria, and other contaminants in public and private water supplies has become commonplace in virtually every state of the Union.

Supply shortfalls, dropping water tables, and increased competition among users for available supplies can be expected to add challenging new dimensions to future safe drinking water programs. Like questions of safety, future problems of availability are not restricted to any one geographic region. Perhaps predictive of trends elsewhere, the supply situation is especially critical in the Western states, where conflicts between the use of water for energy production, agriculture, industrial productivity, and domestic consumption already are visible and worsening.

Given the complex range of questions concerning national priorities, state laws, water rights, and parochial interests, it is unlikely that the task of protecting the adequacy and safety of drinking water supplies will be easily or quickly accomplished. The problems to be resolved are most evident in water-short regions where population growth is causing a decline in the per capita supply of fresh water.

California, faced with one of the nation's fastest growing populations, is a case in point. Already the nation's most populous state, California's Colorado River water allotment is scheduled to be cut by more than 50% as the State of Arizona under a Supreme Court order, preempts a greater share of the Colorado River supply in 1985. Plans to divert fresh water from the north to the southern portion of the state continue to be delayed by a series of legislative, executive, and legal actions, and are further complicated by deficit-conscious public attitudes.

There is a growing realization that in the future water will cost more. Pointing to the rising cost of pumping water from greater depths and the escalating expense of transporting it over increasingly longer distances, some experts have projected that the price of water to California consumers will rise at least 300% within the next ten years. But -- because it fails to allow for any additional water treatment or rejuvenation of aging distribution systems -- others argue that this estimate is too low.

In short, drinking water: where to get it, how to treat it, and how to deliver it seems destined to remain, for the foreseeable future, a priority issue at the national, state, and local levels.

As in the past, we can and must depend on the bottled water industry as an alternate source of safe drinking water whenever (a) natural or man-made emergencies make a water supply unsafe, (b) the cost of bringing a public water supply or distribution system into regulatory compliance is prohibitive, (c) special health needs require low nitrate, low sodium or reduced fluoride diets, (d) a demand exists for drinking water that meets all primary and secondary standards, that is protected in sealed containers against contamination during distribution, and that looks clean and tastes good.

CHAPTER 5

OVERVIEW OF POINT-OF-USE WATER TREATMENT TECHNOLOGY

P. Regunathan, Ph.D.

Everpure, Inc.
660 North Blackhawk Drive
Westmont, Illinois 60559

The treatment of water at the point of use (POU) has increased many fold during the last ten years to meet the real and perceived needs of the consumers in this country. While most POU units are acquired for the improvement in aesthetic quality of the water, more and more devices are becoming available to reduce the concentration of health-related contaminants found in some water supplies. It is an established fact that less than one percent of the water produced by central water treatment plants is actually used for cooking and drinking. This small fraction of the treated water can be further treated economically at the point of use, meeting more stringent requirements than the rest. Some contaminants that are present at the tap, such as distribution system-related taste and odor and corrosion products, products resulting from water main failure, and disinfection by-products, cannot be removed at the central plant. These factors make POU units ideal supplements for producing small quantities of specially treated water for specific needs.

UNITS AND APPLICATION

A variety of POU units exists in this country at this time. They include softeners to remove hardness, adsorption filters for the removal of dissolved organics, particulate filters for removing turbidity, cysts, and asbestos fibers, and reverse osmosis-activated carbon systems to remove dissolved minerals along with the above-mentioned contaminants. Figure 1 shows the basic design of the most common units:

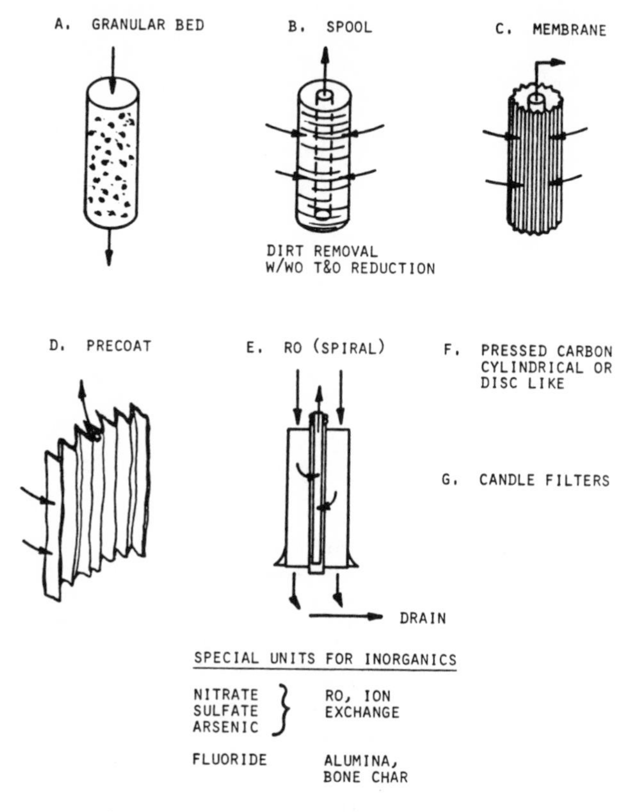

Figure 1. Type of units.

A. Granular Bed -- most common, simplest, and easiest to build. Granular activated carbon beds can be designed to remove a wide variety of organic contaminants.

B. Spool Filters -- made of fibers, string, or rolled paper. Used in a variety of particulate removal function. Efficiency also can vary widely in terms of percent removal and size of particles removed.

C. Pleated Filters -- pleated with paper, non-woven fabric, or membranes with controlled pore size. Applications and variations are similar to spool filters.

D. Precoat Filters -- use finely powdered filter medium, usually diatomaceous earth or filtering material. If properly designed not to channel or crack the filter medium, all particles 1 micron or even smaller can be removed effectively. If activated carbon is in the filter medium, taste and odor removal can be added to the particle removal capabilities.

E. Reverse Osmosis (RO) Units -- pressure forces water through a membrane with Angstrom-sized pores. Accumulated mineral concentration is swept to waste by drain flow. Removes most dissolved substances except volatile organics.

F. Miscellaneous Units -- pressed carbon and candle filters can be designed to combine effective particle removal with organic removal. RO, ion exchange, and activated alumina are used more frequently for the removal of health-related organics.

Application of these different units in the home can be divided into four major categories (see Figure 2):

A. Faucet Add-On -- Mostly carbon bed type, easiest to install, but some are too small for significant reduction of levels of health-related organics.

B. Under-the-Sink Units -- Two subcategories are stationary units which treat all the cold water going to the existing faucet and line bypass units which treat a small portion of cold water diverted to a separate faucet. The RO-carbon system with a reservoir is a special case in this category.

C. Whole Household Units -- other than water used for lawn sprinkling, all the water used in the home is treated by these relatively large units. These are becoming useful in removing volatile organics from contaminated groundwater supply used by private home owners.

D. Unplumbed Units (not shown in figure) -- usually granular carbon bed type. These pour-through units can operate either by gravity or with a pump to treat small quantities of water as required.

The variations in POU units encompass all the different treatment processes that are known to the entire water treatment industry. Types of media used vary in terms of granular or powdered with their inherent size differences, strength and hardness, source of raw materials and the manufacturing process differences. The materials used in the pressure vessels and the internal components

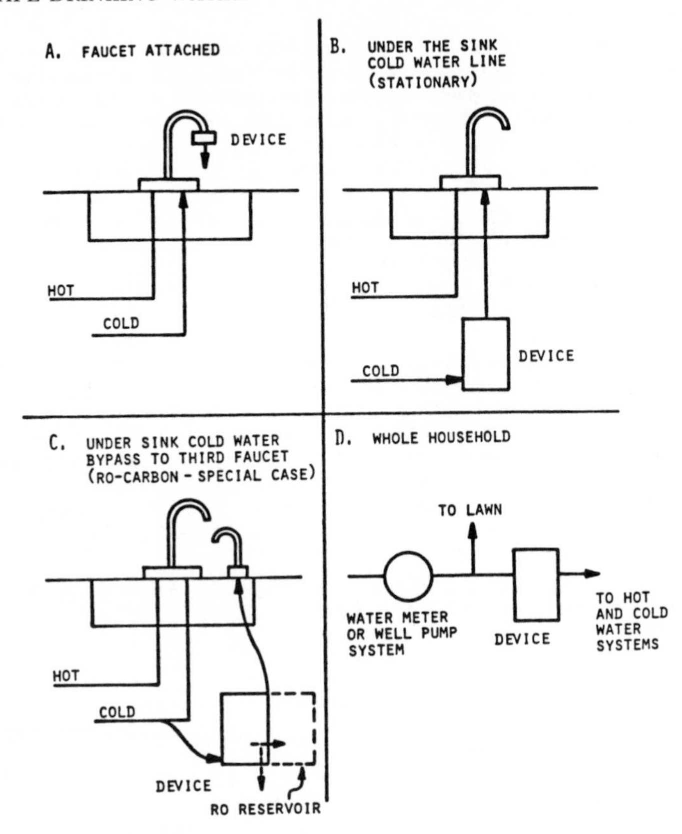

Figure 2. Basic type of system.

can be steel, aluminum, brass, copper, or various plastics or combinations of one or more of the above. Different manufacturers use different means to caution the user to service the units periodically. These means include deliberate understatement of capacity in labelling, use of meters or pressure-sensing controls, or a combination of the above. Some units require removal of cartridges inside permanent pressure vessels; others have disposable pressure vessels with their own incorporated cartridges; some other types of units require in-place back-washing rather than change of media.

CHEMICAL REMOVAL

The original need that is still predominantly required from the point of use units is that of taste and odor removal along with visible dust and rust removal. However, health-related organic and inorganic removal is gaining importance in the industry, as more and more potentially harmful contaminants are being detected in small public and private water systems. While the POU manufacturers have carried out their own performance studies, the most comprehensive and independent study of these units for their effectiveness in the removal of volatile organics was completed by Gulf South Research Institute (GSRI) under a program funded by the U.S. Environmental Protection Agency. Thirty different units of various design and manufacture were tested under this program for their effectiveness in removing trihalomethanes, and various other volatile organics (1,2,3,4). At least ten POU units were found to be effective in that they removed higher than 85% of the volatile organics for their rated life in these tests.

POU units containing activated carbon have been of practical use in several locations across the continental United States to remove or reduce levels of various volatile organics found to be present in the water supply. A partial list of the organics, their concentrations and the locations is shown in Table I. Most of the organics are synthetic materials, probably leached from landfills or similar sources. Trichloroethylene-type compounds are the most often found. POU units used in these locations reduced the concentrations shown in the table to a very low range, often below detectable levels for the particular organic compound.

The POU industry's own in-house tests include organic and inorganic removal by various units, such as activated carbon units, RO-carbon systems, softeners and activated alumina systems. Several examples of such work and the data will be presented in this section.

A dual cartridge system with a granular carbon bed and a precoat filter was tested for the removal of chloroform (average 320 microg/L) and carbon tetrachloride (27 microg/L) spiked into a municipal water with a natural average TOC concentration of 5.0 mg/L. The test conditions and the results have been described in detail elsewhere (5), but are shown briefly in Figure 3. This dual cartridge system was found to be capable of reducing the levels of both compounds by more than 80-90 percent for at least 1,000 gallons in this test. The same system also has been tested by GSRI in their comprehensive test program (1,2,3), indicating a removal capability of more than 90% for TTHMs, 40-50% of TOC, approximately 80% of NPTOX, and more than 95% of other volatile organics listed in Table II.

An RO-carbon system operable on line pressure in a home has been tested for the removal of various organics, inorganics, and particulates. A deep well water, softened by ion exchange and spiked with various contaminants and humic acid was used in several separate tests to gather the data shown in Table III. With the exception of silver and

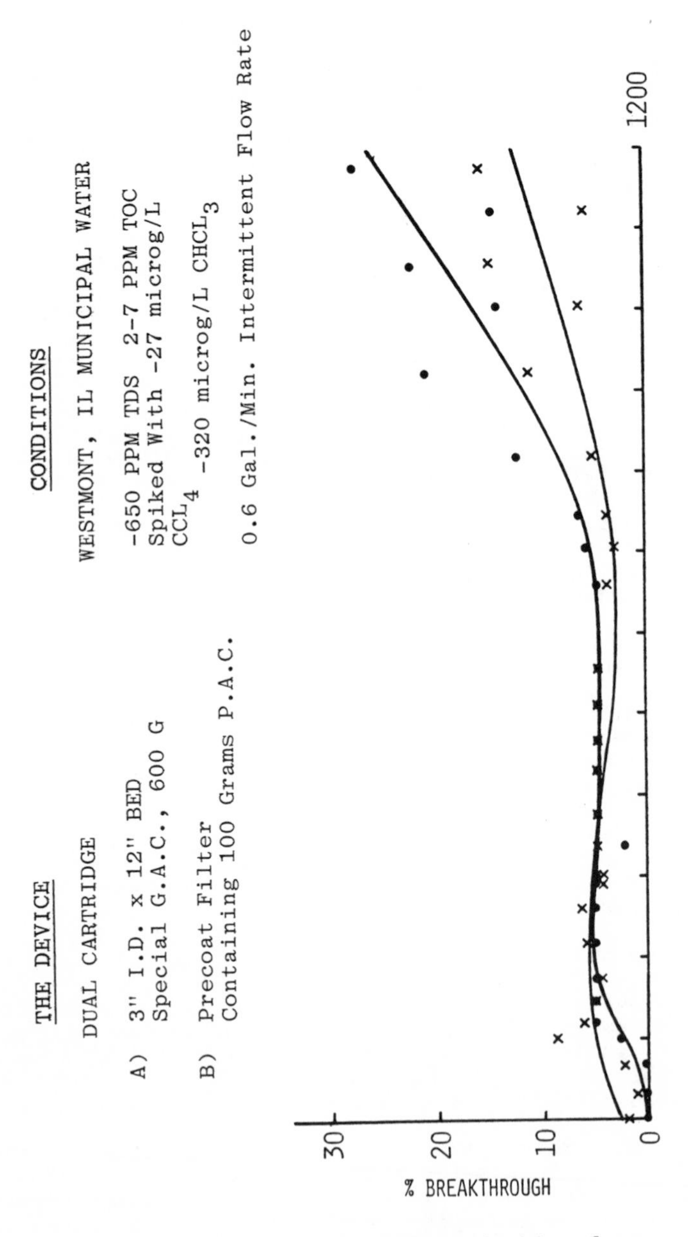

Figure 3. Total gallons filtered-chloroform and carbon tetrachloride reduction by a P.O.U. device.

TABLE I. PRACTICAL USE OF POU UNITS

COMPOUND	INFLUENT LEVEL, PPB	LOCATION
DBCP	3 - 5	KINGSBURG, CA
	3 - 5	REEDLEY, CA
TRICHLOROETHYLENE	70 - 290	NEW BRIGHTON, MN
	90 - 240	ROCKAWAY, NJ
	20 - 100	SILVERDALE, PA
1,1,1-TRICHLOROETHANE	50 - 240	ROCKAWAY, NJ
	UP TO 1000	MIDSTATE, CT
ALDICARB CARBAMATE	20 - 515	SUFFOLK COUNTY, NY
1,1,2-TRICHLOROETHYLENE	5 - 20	ROCKAWAY, NJ
1,1,1-TRICHLOROETHANE	1 - 10	ROCKAWAY, NJ
1,2-DICHLOROPROPANE	20 - 515	ROCKAWAY, NJ
TETRACHLOROETHYLENE	1 - 20	ROCKAWAY, NJ
	10 - 140	SILVERDALE, PA
CHLORDANE	1 - 10	PITTCAIRN, PA
THM'S	100 - 500	SEVERAL LOCATIONS
PHENOL (SPILL)	4 - 7 PPM	NEW ORLEANS, LA

TABLE II. GSRI TESTS

COMPOUNDS	INFLUENT LEVEL, PPB	LOCATION
Trihalomethanes Chloroform Bromoform Chlorodibromomethane Bromodichloromethane	9 - 300	New Orleans, LA Miami Beach, FL Atlanta, GA Detroit, MI Pico Rivera, CA
Trichloroethylene	50	Spiked Ground Water and Spiked New Orleans, LA Water
Tetrachloroethylene	50	
Carbon Tetrachloride	20	
1,1,1-Trichloroethane	50	
Chlordane	50	Spiked New Orleans, LA Water
p-Dichlorobenzene	10	
Hexachlorobenzene	10	

TABLE III. SUMMARY OF TEST RESULTS* OF AN RO-CARBON DEVICE**

CONTAMINANT	INFLUENT CONCENTRATION	PERCENT REMOVAL
TOTAL DISSOLVED SOLIDS, *mg/L*	1275	88
TOTAL ORGANIC CARBON, *mg/L*	11	>90
THM, *μg/L*	770	>90
CARBON TETRACHLORIDE, *μg/L*	20	>95
ENDRIN, *μg/L*	2	>99
METHOXYCHLOR, *μg/L*	1000	>99
LINDANE, *μg/L*	40	>99
PCB, *μg/L*	100	>99
NITRATE, *mg/L*	100	40
FLUORIDE, *mg/L*	100	85
CHLORIDE, *mg/L*	470	87
SULFATE, *mg/L*	215	98
SODIUM, *mg/L*	270	82
CHROMIUM, *μg/L*	3400	88
CADMIUM, *μg/L*	900	76
BARIUM, *mg/L*	10	71
LEAD, *μg/L*	2100	72
SILVER, *μg/L*	600	34
ASBESTOS FIBERS 10^6 *fibers/L*	>200	>99

* TRACE METALS, PCB, ASBESTOS FIBER TESTS WERE SHORT DURATION TESTS. OTHERS TESTED FOR 1000 GALLONS. TEST PRESSURE -50 PSI

** REVERSE OSMOSIS MEMBRANE SYSTEMS FOLLOWED A BED WITH 275 GRAMS OF CARBON.

nitrate, the removal rates ranged from 71% for barium to 99% for organic pesticides and asbestos fibers. Removal rates for silver and nitrate were 34% and 40%, respectively, in these tests. More recent unpublished data show that these rates can be improved further by the use of improved membranes or higher influent pressure.

Cadmium and hardness removal by a home water softener was studied in one test and the results obtained during the 527th regeneration cycle are shown in Figure 4. Influent cadmium concentration and the hardness levels were 1.0 mg/L and 370 mg/L (as $CaCO_3$), respectively. This length of

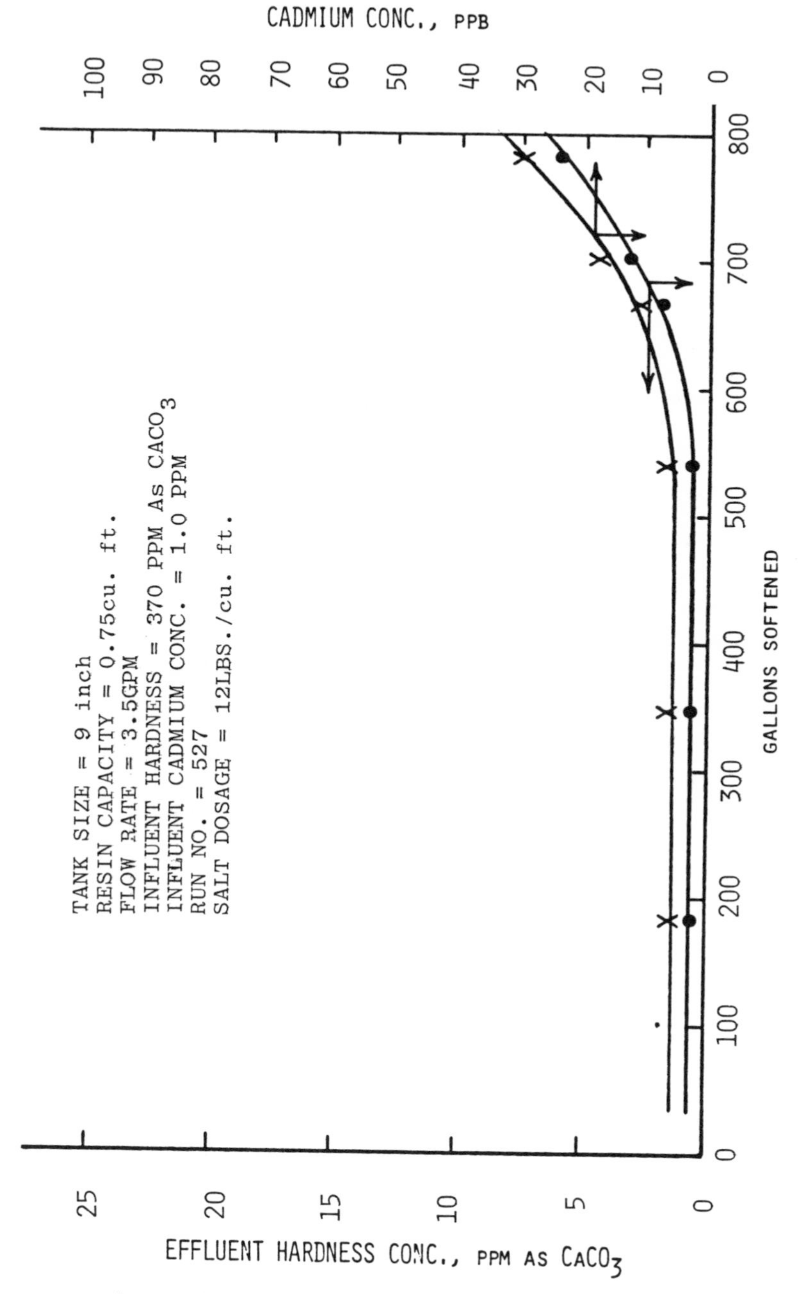

Figure 4. Cadmium removal by a home water softener.

operation (527 cycles) means approximately three years of normal use of the water softener in an "average" home. The removal curve for cadmium closely resembles that of hardness, implying that normal regeneration by salt is sufficient for continuing removal of cadmium by softening resins.

The heavy metal removal abilities of a home water softener using cation exchange resin in sodium form along with that for cadmium are summarized in Table IV. This table shows that while cadmium removal can be restored to the same level after successive regenerations, lead, barium, chromium, and silver tend to accumulate as the operation is continued from cycle to cycle. However, even after 527 cycles, lead removal was 92% and after 192 cycles, the other three metals were removed by more than 90%. This indicates that if a home water softener is to be used for the removal of these four heavy metals (other than cadmium), yearly cleaning of the resins with acids and salt would be necessary to restore them to virgin conditions for continued heavy metal removal.

A special two-cartridge system with activated alumina has been operated to establish their capability to remove pentavalent arsenic levels found in some groundwater supplies. Each of the two cartridges was 2.5 in. I.D. x 10 in. deep, filled with activated alumina. The system was operated continuously at 0.5 gpm with influent arsenic(V) in concentrations of 0.87 ± 0.14 mg/L. The breakthrough after the first cartridge and the system effluent are shown in Figure 5, indicating the system to be effective in maintaining the arsenic(V) level below the MCL of 0.05 mg/L for 700 gallons of water.

TABLE IV. HEAVY METAL REMOVAL BY CATION EXCHANGE RESIN

HEAVY METAL	INFLUENT CONC., PPB	NUMBER OF TEST CYCLES	EFFLUENT CONC., PPB		PERCENT REMOVAL LAST CYCLE
			DURING EARLY CYCLES	LAST CYCLE	
CADMIUM	1,000	527	<10	<10	>99
LEAD	1,000	527	10	80	92
BARIUM*	10,000	192	10	1,000	90
CHROMIUM*	1,000	192	<10	20	98
SILVER*	1,000	192	10	80	92

* 2-INCH DIAMETER COLUMN TESTS (SIMULTANEOUSLY TESTING ALL THREE METALS ALONG WITH NATURAL HARDNESS IN THE WATER)

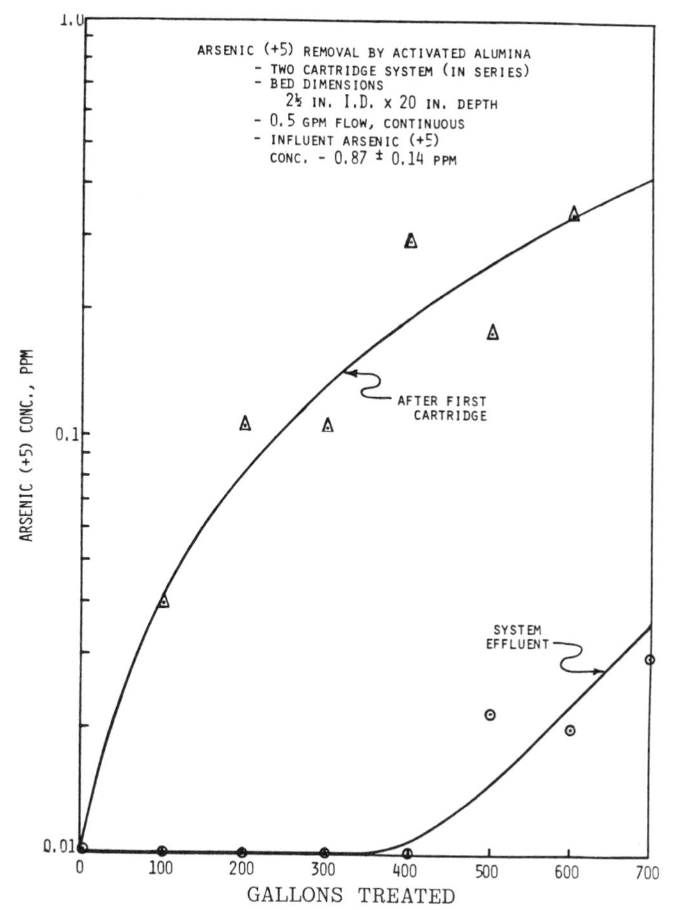

Figure 5. Arsenic (V) removal by activated alumina.

These examples of chemical removal capabilities of different types of POU devices indicate a wide ranging ability to deal with chemical contaminants at the point of use. Thus, often when construction costs and time tables are unattractive, installation of POU devices can be a better solution to water quality problems than continued reliance on a central treatment system.

PARTICULATE REMOVAL

POU products designed for particulate removal vary widely in their effectiveness due to the fact that different applications, including industrial and commercial needs, require different efficiencies and life capabilities. Some of these products are merely strainers removing only large particles that are visible to the naked eye. Fine filters do exist that can remove 99.9% or more of micron or submicron sized particles. There is some confusion about the "nominal" rating of filters in terms of particle removal, due to the varying interpretations by manufacturers and users. The recently developed National Sanitation Foundation Standard 42 defines the "nominal" rating as the smallest particle size at and above which 85% of all particles would be removed by the filter (6). Even this is a compromise definition. So, caution is required in using these products in specific applications with specific needs.

Particulate contaminants of established or potential health-related concern include turbidity, asbestos fibers, protozoan cysts, particles that have adsorbed on their surfaces heavy metals and/or organics, and heavy metal precipitates. Among these different types of particulates, the cysts of Giardia lamblia are gaining increasing importance as the Giardia outbreaks show a doubling every five years (7).

Some POU products have been tested for their abilities to remove fine particles. One such test using precoat filters studied the influent and effluent particle counts using a volcanic ash suspension with the particle size ranging from 1.0 to 60 microns (5). A removal ability of 99.9% of all particles was attained by this precoat filter (see Figure 6). The same type of precoat filter was tested at Duluth, Minnesota with an average influent amphibole asbestos fiber concentration of 120 MFL (8). The filter reduced the fiber concentration to below detectable levels during this test lasting 3,000 gallons of water, establishing a removal of greater than 99%.

A recent study investigated seven POU devices for their Giardia lamblia cyst removal effectiveness (9). Four of the seven tested removed 100% of the cysts introduced into the influent test water, while the other three reduced the cyst level by 66.4 to 84.4%. The results of this test are summarized in Table V.

BACTERIOLOGICAL CONCERNS

Many have expressed concerns about the possibility of excessive bacterial growth in activated carbon and other POU devices due to intermittent or extended use of the products by the consumers. Indigenous bacteria in a water supply can grow to considerable numbers under quiescent conditions. This is basically due to the removal or dissipation of disinfectant residuals in the presence of nutrients and the surface area presented in the system. Such growth can occur under certain conditions on the walls of pipes, valves, fittings, faucets, as well as on the surface of the filtering media in POU products. Activated

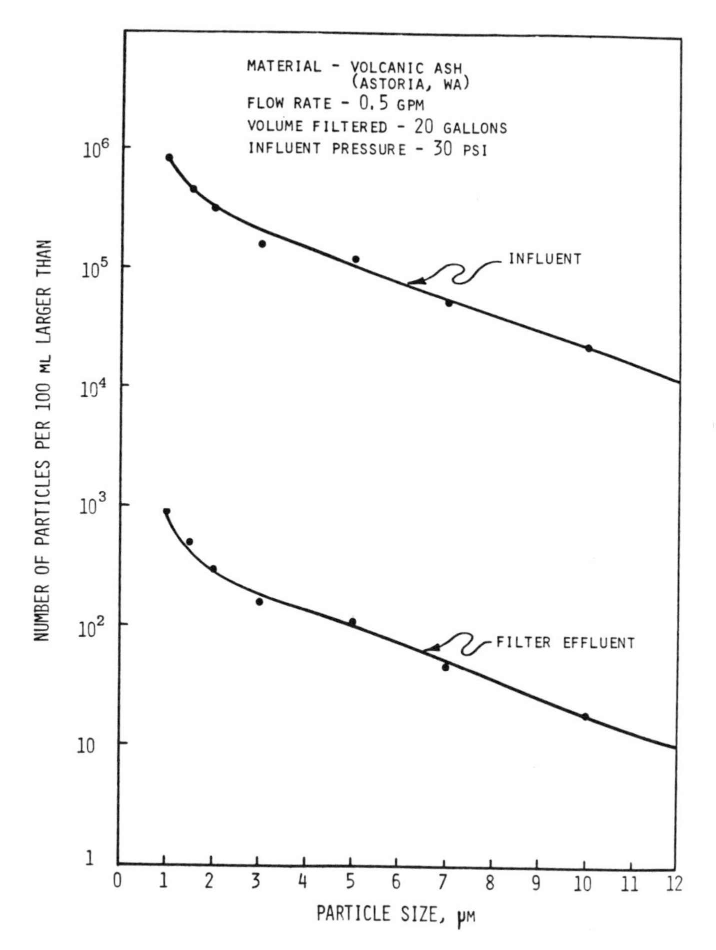

Figure 6. Particulate removal by a precoat carbon filter.

carbon or other filter media are not unique in this aspect. All surfaces can support such growth. Even RO-carbon systems and "fine" filters allow a few organisms to pass through and colonize on the effluent side of the systems.

The extent and level of such growths in POU products was studied extensively in the GSRI study (1,2,3). A summary of their results in Phase 2 is shown in Table VI. A comparison of the mean values in the influent (New Orleans water) and the effluent from activated carbon POU devices show that a one-half to one log increase of

TABLE V. CYST REMOVAL BY POU PRODUCTS (9)

FILTER	CYSTS INTRODUCED	CYSTS RECOVERED	% EFFICIENCY
Quality Control #1	50,000*	48,600*	97.2%*
Everpure Model QC4-SC	50,000	0	100.0%
Pollenex Model WP-100	50,000	16,800	66.4%
Sea-Gull IV Model X-IF	50,000	0	100.0%
Micro-Twin Model 10-TOBC	50,000	0	100.0%
Mr. Water Filter Model SY-1	50,000	15,000	70.0%
Royal Doulton Model F303	50,000	0	100.0%
Water-Pik Instapure Model F-1C	50,000	7,800	84.4%
Quality Control #2	50,000*	48,800*	97.6%

* Since the counts were not absolute numbers, the recovery of 97.2% and 97.6%, respectively, represents as closely as possible our recovery techniques.

TABLE VI. GSRI (PHASE 2) DATA AND STATISTICAL COMPARISONS

		MEAN LOG_{10} VALUES		
	RUNNING	OVERNIGHT	2 DAYS OFF	5 DAYS OFF
SPC 1* INFLUENT	1.42	1.61	2.32	2.68
SPC 1 EFFLUENT	2.18	2.69	3.04	3.12
F TEST 95% CONF. LEVEL	NO	YES	YES	YES
T TEST 95% CONF. LEVEL	NA	YES	YES	YES
SPC 2** INFLUENT	1.47	1.72	2.32	2.93
SPC 2 EFFLUENT	2.31	2.75	3.02	3.13
F TEST 95% CONF. LEVEL	YES	YES	YES	YES
T TEST 95% CONF. LEVEL	YES	YES	YES	NO

* - 35^{0} C FOR 48 HOURS

** - 25^{0} C FOR 120 HOURS

INDUSTRY'S OWN CONCLUSION

- NO HEALTH RELATED DIFFERENCE BETWEEN THE INFLUENT AND EFFLUENT VALUES heterotrophic bacterial count can occur due to the presence of these devices. Since the actual counts in the influent waters often can be as high or higher than the effluent levels, it is reasonable to conclude that a health-related concern due to heterotrophic bacterial growth need not be maintained.

Even though the POU industry has been active in attempting to incorporate a bacteriostat in these POU devices in order to remove this as a factor of focus, none yet has been found to be effective. This is due to the resistance of varied species of organisms that are found in different supplies across the country, water quality variations, and a fairly high input of these organisms into the units. The only bacteriostat that has been tried by many is silver deposited on the activated carbon media. It was shown in the GSRI study that silver-containing carbon POU units did not control bacterial growths any better than non-silver units in New Orleans (4).

However, silver does have some measurable effect on some specific organisms when allowed to act slowly. In one study, silver-containing powdered carbon precoat filters were compared with similar control units without silver in reducing the levels of *Ent. aerogenes* introduced into the influent water. The organism level in the influent was maintained at 1,000 organisms per 100 mL during the first week of tests, and then increased to 100,000 organisms per 100 mL during the last two weeks of tests. Control filters reduced these levels by fine filtration to 10 organisms per 100 mL during the first week and 100 to 1,000 organisms per 100 mL during the last two weeks, indicating a fairly uniform removal rate of 99% by fine filtration. Filters with silver reduced these influent levels even further to <1 to 1 organism per 100 mL, and later to approximately 10 organisms per 100 mL. The results shown in Figure 7 indicate the effect of silver to be 1 to 1.5 log on this coliform organism. Simultaneous analysis of heterotrophic bacterial counts showed no effect on these organisms by silver, as shown in Figure 8. It can be concluded from these tests that silver may be effective specifically against certain organisms, but not against all organisms -- at least in certain water supplies.

The incorporation of an economically acceptable antimicrobial agent in POU devices has not yet become technologically feasible due to varied resistance of different microbial species to a particular antimicrobial agent under varying water quality conditions and due to the occasionally high input of bacteria into a POU product. Since no health-related difference exists between the bacterial growth in distribution systems and in POU products, this can only be of academic interest, not an issue of practical importance.

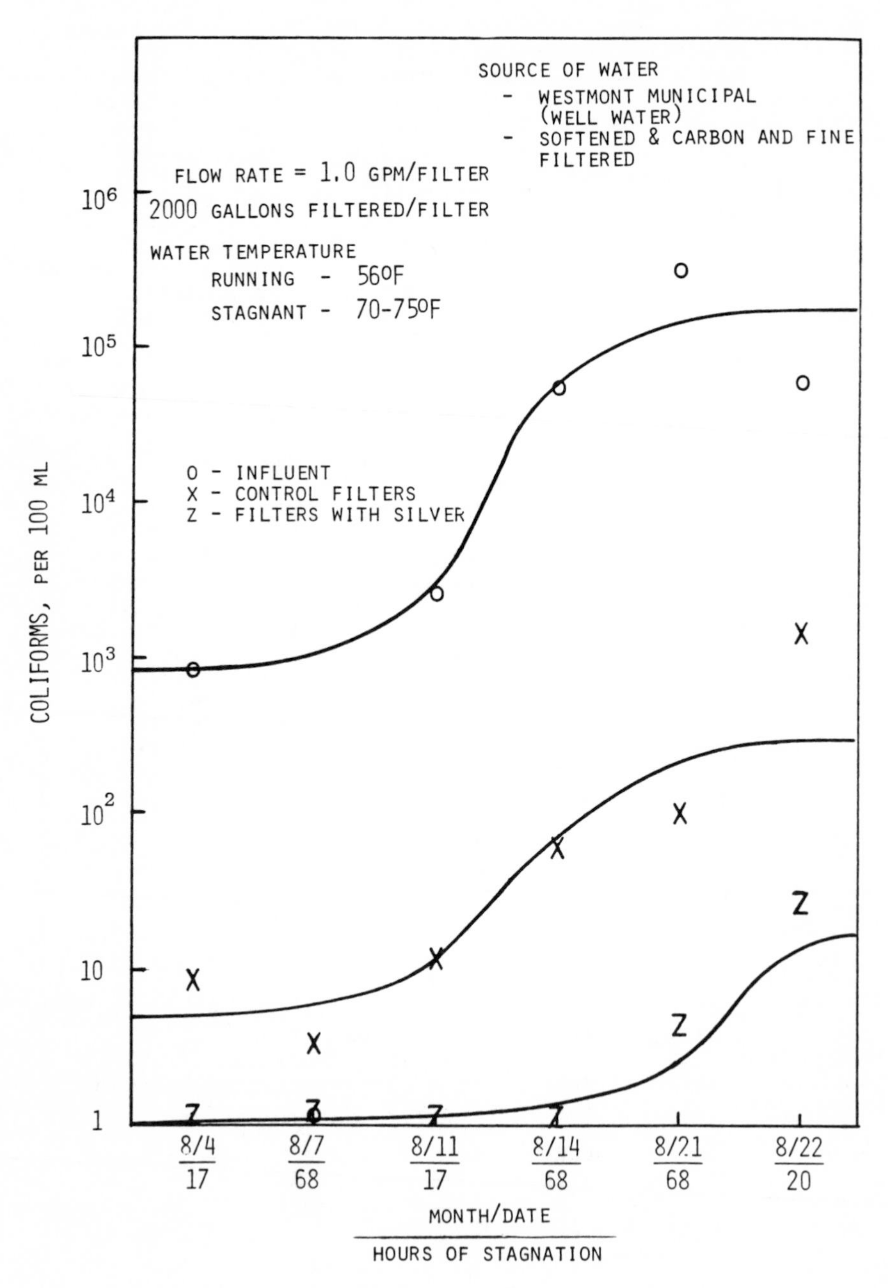

Figure 7. Effect of silver on coliform.

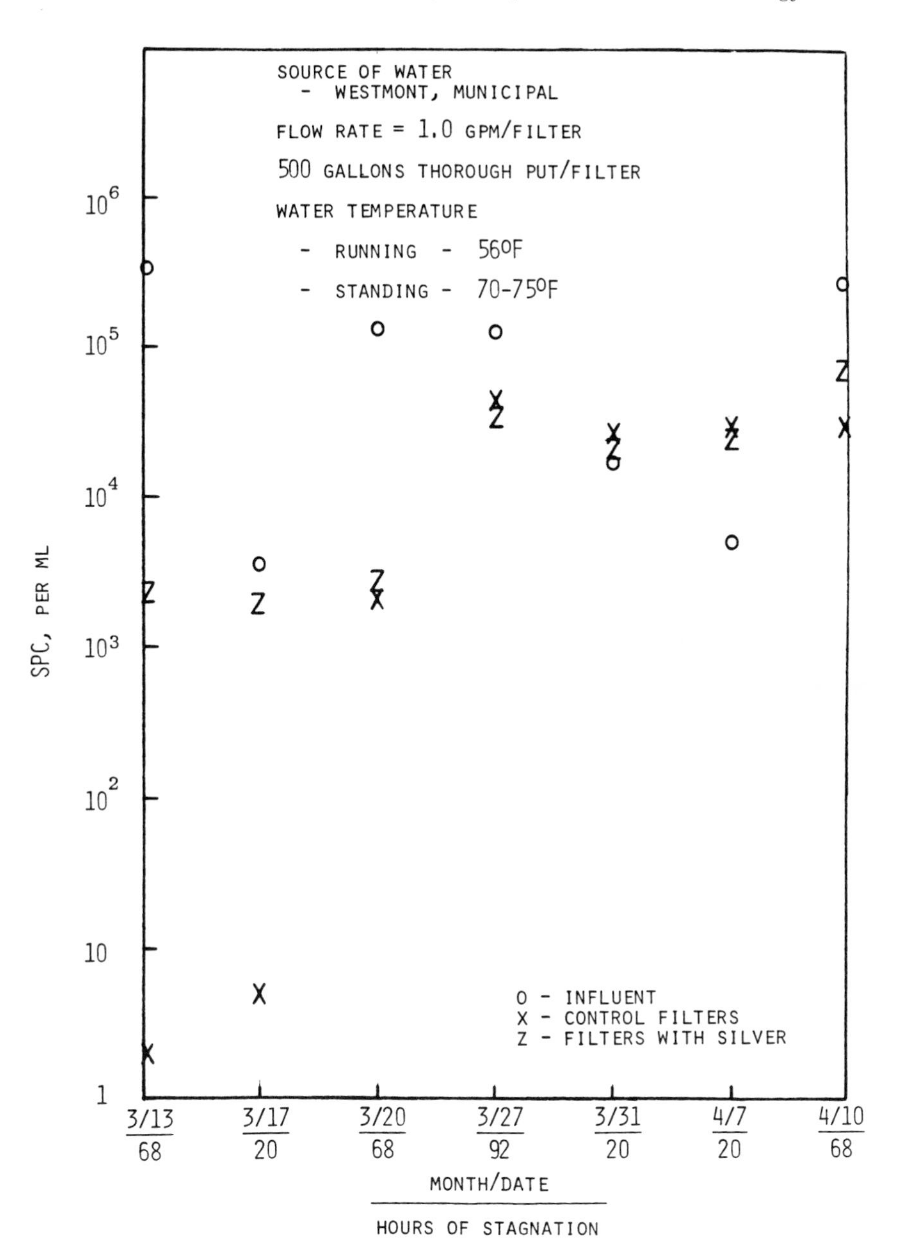

Figure 8. Effect of silver on SPC.

COST FACTORS

Recently, a study was completed by Temple, Barker, and Sloane (TBS) of Lexington, MA for the Water Quality Association concerning the cost of POU treatment compared to central treatment (10). In this study, six basic processes were considered; reverse osmosis, cation exchange, anion exchange, activated alumina, direct filtration, and granular activated carbon filtration. These POU products were considered to treat contaminants that may cause chronic health effects. The scenarios considered were drinking water only (separate faucet), all the cold water used in a single tap (stationary), and whole house. The results showed the following (10):

POU Treatment	Cost-Competitive Community Size	Amount of Water Treated
Reverse Osmosis	All sizes	Drinking water only
	<90 houses	Single tap
Cation Exchange	All sizes	Drinking water only
	<50 houses	Single tap
	<30 houses	Whole house
Anion Exchange	<85 houses	Drinking water only
Activated Alumina	<100 houses	Drinking water only
	<80 houses	Single tap
	<60 houses	Whole house
Direct Filtration	All sizes	All scenarios
Granular Activated Carbon	All sizes	All scenarios

The treatment of drinking water only with a separate faucet was judged to be lower in cost than central treatment in all sizes of communities when considering four of the six processes: reverse osmosis, cation exchange, direct filtration and granular activated carbon filtration. This study has shown the economic viability of the POU approach.

CONCLUSIONS

1. Properly designed products can remove a wide variety of contaminants including heavy metals, THMs, other organic contaminants, particulates, and in some cases microorganisms. These also can result in drinking water with improved taste, odor, and clarity.

2. The point-of-use products, if properly designed, can be cost-effective supplements to centralized water treatment systems and can be beneficial in locations where the utilities cannot be cost-effective due to their size or other economic factors. They can be cost-effective alternatives to central treatment for specific, chronic contaminants in smaller communities.

3. Industry now has growing capabilities in the development of increasingly more effective point-of-use water treatment products.

REFERENCES

1. Smith, J.K., et al. "Development of Basic Data and Knowledge Regarding Organic Removal Capabilities of Commercially Available Home Water Treatment Units Utilizing Activated Carbon, Preliminary Report -- Phase 1," Gulf South Res. Inst., New Orleans, LA. (May 1979).
2. Perry, D.L., Smith, J.K. & Lynch, S.C., "Development of Basic Data and Knowledge Regarding Organic Removal Capabilities of Commercially Available Home Water Treatment Units Utilizing Activated Carbon, Final Report -- Phase 2," Gulf South Res. Inst., New Orleans, LA. (July 1980).
3. Smith, J.K., et al., "Development of Basic Data and Knowledge Regarding Organic Removal Capabilities of Commercially Available Home Water Treatment Units Utilizing Activated Carbon, Preliminary Report -- Phase 3 (Final)," Gulf South Res. Inst., New Orleans, LA. (October (1981).
4. Bell, F.A., Jr., et al., "Studies on Home Water Treatment Systems," J. Am. Water Works Assoc. 76(4):126 (1984).
5. Regunathan, P., et al., "Efficiency of Point-of-Use Treatment Devices," J. Am. Water Works Assoc. 75(1):42 (1983).
6. National Sanitation Foundation, "Drinking Water Treatment Units -- Aesthetic Effects, Standard Number 42," Ann Arbor, MI (June 1982).
7. "Giardiasis Outbreaks Show Steady Increase," Mainstream, Am. Water Works Assoc. 278(7):1 (1984).
8. Cook, P.M., et al., "Evaluation of Cartridge Filters for Removal of Small Fibers from Drinking Water," J. Am. Water Works Assoc. 70(8):459 (1978).
9. Hibler, C.P., "An Evaluation of Filters in the Removal of *Giardia lambia*," Water Technology 7(4):34 (1984).

10. Temple, Barker & Sloane, Inc., "Point-of-Use Treatment for Compliance with Drinking Water Standards," Lexington, MA (May 1983).

CHAPTER 6

AN EFFECTIVE ALTERNATIVE TO OFFICIAL REGULATION OF INDIRECT ADDITIVES TO DRINKING WATER

Nina I. McClelland, Ph.D.

President
National Sanitation Foundation
Ann Arbor, Michigan 48105

DEFINITION

By definition, additives to drinking water may be grouped into two broad general classes: _direct_, chemicals which are added intentionally for the treatment of water; and _indirect_, chemicals which are added unintentionally by contact of the water with various materials and items of equipment.

REGULATION

The U.S. Environmental Protection Agency (EPA) has mandated, primary authority for regulating additives and other substances in drinking water under Public Law (PL) 92-523 (the Safe Drinking Water Act - SDWA), PL 94-469 (the Toxic Substances Control ACT - TSCA), and PL 92-516 (the Federal Insecticide, Fungicide, and Rodenticide Act - FIFRA). This authority was clarified in an EPA and Food and Drug Administration (FDA) Memorandum of Understanding (MOU), published on July 20, 1979 (1). In accordance with the MOU, FDA has authority under the Federal Food, Drug and Cosmetic Act (FFDCA) for regulating bottled water, and water used in food and food processing facilities; EPA has authority for regulating the quality of all other drinking water covered by Federal regulations.

Under the SDWA, EPA's authority is limited to public water supplies - the 59,660 community systems serving 25 or more persons, and 160,000 non-community systems serving the traveling public. Private wells and small cluster systems serving less than 25 people are not regulated under the Act.

The World Health Organization (WHO) has in final draft form, Guidelines for Drinking Water Quality, a document intended for use in developing standards or alternative control procedures for all drinking water, including wells and bottled water (except bottled mineral water, which is considered a beverage). This Guidelines document supersedes both the WHO European (1970) and International (1971) Standards for Drinking Water. It includes inorganic and organic chemicals, microbiological and biological contaminants, radionuclides, and aesthetic parameters.

In the U.S. the Public Water Supply Program is a strong partnership in which EPA sets national standards for contaminants which "may have any adverse effect on the public health" (SDWA Section 1412), and the states or territories elect to accept primary enforcement responsibility. (There are 57 eligible states and territories: to date, 51 have accepted "primacy" under the statute).

As of this writing, EPA has established primary standards - maximum contaminant levels (MCLs) - for ten inorganic chemicals, six pesticides, total trihalomethanes, radionuclides, coliform bacteria, and turbidity. MCLs are established for long term ingestion of the contaminant; i.e., they assume a daily consumption per individual of two liters of drinking water and 6.5 g of fish products over a lifetime of 70 years. Fifteen volatile organic chemicals (VOCs) are being considered for regulation (2).

In addition to its standard setting functions, EPA is granted emergency powers under SDWA (Section 1431), and provides technical guidance through its program of Health Advisories [HAs, vis-a-vis SNARLs (Suggested No Adverse Response Levels)]. Twenty-three HAs have been developed to date on a need basis to assist with contamination incidents; e.g., accidental spills and groundwater problems. In setting standards (MCLs), health effects, generally available treatment technologies, and costs must be considered. By contrast, only health effects are considered in the HAs, which are not Federally enforceable standards. Current U.S. standards, WHO Guidelines, and other national standards are summarized in Table I.

To deal with the issue of additives, EPA established an Additives Evaluation Branch in 1980. Through the Branch, a contract was awarded to the National Academy of Sciences (NAS) to develop and publish a Water Chemicals Codex for direct additives, and advisory committees were established to assist in dealing with indirect additives. The first edition of the Codex was published in 1982 (3) and is the subject of another paper in this symposium (4). To date, EPA has provided technical assistance to states in the form of "advisory opinions" regarding additives, but it has not "approved" or otherwise controlled either direct or indirect additives.

TABLE I. MAXIMUM LIMIT VALUES FOR HEALTH RELATED INORGANIC SUBSTANCES[a] VARIOUS COUNTRIES

	PARAMETER							
	Sb	As	Ba	Cd	Cr	Pb	Hg	Se
USA 1976 Maximum Contaminant Level	-	0.05	1.0	0.01	0.05	0.05	0.002	0.01
WHO Proposed 1983 Guideline Value	-	0.05	-	0.005	0.05	0.05	0.001	0.01
WHO (European) 1970 Upper Limit Concenatration	-	0.05	-	0.01	0.05	0.1	-	0.01
WHO (International) 1971 Upper Limit Concentration	-	0.05	-	0.01	-	0.1	0.001	0.01
EEC 1980 Maximum Admissible Concentration	0.01	0.05	-	0.005	0.05	0.05/ 0.1	0.001	0.01
USSR 1972 Maximum Permissible Concentration	0.05	0.05	4.0[b]	0.01	0.1/ 0.5[b]	0.1	0.005[b]	0.001
CANADA 1978 Maximum Acceptable Concentration	-	0.05	1.0	0.005	0.05	0.05	0.001	0.01
INDIA 1976 Cause for Rejection	-	0.05	-	0.01	0.05	0.1	0.001	0.01
EGYPT 1980 Maximum Permissible Limit	-	0.05	-	0.01	-	0.01	0.001	0.01
JAPAN 1978 Maximum	-	0.05	-	0.01	0.05	0.1	0.005	0.01
AUSTRALIA 1980 Health Investigation Levels	-	0.05	1.0	0.01	0.05	0.05	0.001	0.01

[a]All values are mg/L

[b]Value footnoted, but footnote not identified in available reference.

INDIRECT ADDITIVES: PRODUCTS AND ISSUES

Products which may contribute indirect additives to drinking water include pipes, fittings, and appurtenances used as components of distribution and plumbing systems; paints; coatings; joining materials, including caulks and sealants; films and liners; and other items which are in direct contact with the water at any stage prior to access by the consumer. There are hundreds - perhaps thousands - of these products in common use.

The principal issue is adverse health effects attributed to leaching of inorganic and organic chemical contaminants into the water; i.e., indirect additives. Recently, permeability of piping system components also has become a major issue.

It is important to recognize that standard setting is a complex process, which requires both sound science and good judgement. We are addressing today chemicals assumed for decades to be either safe or absent from our water supplies; products which were a way of life are suddenly suspect. The state of our understanding and our ability to deal with drinking water contaminants is changing rapidly.

To identify contaminants which are candidates for regulation, national monitoring surveys have been conducted. To measure the contaminants, analytical techniques - and, in some cases, instrumentation - have been developed and validated. Data to establish toxicity and carcinogenicity have been acquired through animal testing, and occasionally, human experience. The art of science of risk assessment is being developed in a closely parallel manner.

There is a reference to toxicity and adverse health effects from metal piping systems - copper, galvanized, and lead (5-7). Herrera et al (8,9) published theoretical and experimental data supporting the use of tin/antimony (in lieu of lead-based) solders to reduce the levels of metals leached into drinking water from copper piping systems. Experience with asbestos and tetrachloroethylene from asbestos-cement (A/C) and lined A/C pipe is well documented.

Plastics have been used in the U.S. for piping system components since the early 1950s. The types used for potable water (PW) include principally, polyvinyl chloride) (PVC), chlorinated polyvinyl chloride (CPVC), a polyethylene (PE), and polybutylene (PB) . Acrylonitrile-butadiene-styrene (ABS) pipe fittings are used principally for drainage systems but occasionally, for PW. Alleged leaching of organic chemicals from plastics piping system components, including joining materials, is being debated in several state and local health and plumbing code jurisdictions, including but not limited to: California, Iowa, and Michigan.

In California, a 1977 proposal to permit expanded use of plastic plumbing system components in residential construction has resulted in extensive political and technical activity. In 1978, following oral and written

---

[a]May be used for hot or cold water applications.

public testimony, the California Commission of Housing and Community Development (CHCD) filed an administrative action stating that no significant adverse environmental impacts were expected from the use of plastic pipe, and no Environmental Impact Report (EIR) was required.

In January 1980, Assemblyman Papan introduced into the California Legislature, Assembly Concurrent Resolution No. 98 (ACR 98) (10) directing CHCD to hold its decision on expanded use of plastics until a comprehensive special study report was available from the Toxic Substances Alert System of the State Department of Health Services (HALTS). The report was to include, but not be limited to, "a consideration of the safety of" acrylonitrile (AN), methyl ethyl ketone (MEK), methyl butyl ketone (MBK), N,N-dimethylformamide (DMF), cyclohexanone, tetrahydrofuran (THF), and polyvinyl chloride, "as well as any other chemicals found in the cement and primers used for the installation of plastic pipes". ACR 98 also required a comprehensive report from the State Fire Marshall on the potential flammability of plastic pipe and the fire hazards associated with its use.

A comprehensive, short-term study was organized, managed by HALTS, and supported by manufacturers of plastic pipe. Two end use simulations were included in the study: a static system, typical of new construction; and a flow-through system, typical of post-occupancy usage. The systems included pipe, fittings, and "good" and "bad" solvent welded joints. In addition to the solvents in joining materials, organic and inorganic chemicals were analyzed in the extractant waters. Results were reported as follows (11):

- mg/L levels of solvents from cements and primers were found in water residing for two weeks in PVC and CPVC pipes; the levels decreased rapidly after periodic refills. After 98 days of contact, MEK and THF were measured at levels of 1-3 microg/L, and no other compounds were detected.

- higher levels of solvents leached from joints with excess primer than from joints prepared in accordance with manufacturer's instructions.

- Solvent levels decreased rapidly during the early flow period in the flow through (occupancy use) simulation.

---

[a]CHCD was established in 1971 to assist the legislature and provide a public forum for housing issues. The Department of Housing and Community Development (DHCDE) has authority under the State Housing Law to control plumbing standards in California.

o microg/L levels of chloroform and carbon tetrachloride leached from CPVC but not from similar PVC test systems.

o vinyl chloride monomer (VCM) was not detected in any samples.

HALTS reviewed the James M. Montgomery report (11), and concluded that, "with the adequate flushing of new systems prior to occupancy, accumulation to toxic levels of solvents in systems using plastics is unlikely." A procedure for flushing new systems - both plastics and metals - was recommended.

Subsequently, however, CHCD voted to require an EIR, and in March 1983, an Environmental Review Document (ERD) (12) was published which reviewed and interpreted available data, and identified areas of additional need. The summary concludes with, "no clear environmental preference between plastic and metal pipes has emerged from lack of information. Even after the recommended testing, residual uncertainties will exist, and CHCD will need to balance the possible residual risks against the benefits of allowing more flexibility in the choice of plumbing materials."

In February 1982, EPA published a project summary report "Organic and Organotin Compounds Leached from PVC and CPVC Pipe" (13). The purpose of this study was to demonstrate whether or not organic species of tin-based heat stabilizers used in PVC and CPVC pipe leached into water in contact with the pipe. The results: alkyltin species were extracted from PVC at levels of 35 ppb for day one, and 3.0 to 0.25 ppb per 24 hours (biphasic) for days two through 22. For CPVC, the levels were 2.6 ppb for day one, and 1.0 to 0.03 ppb per 24 hours (biphasic) for days two through 21. The question of whether or not these levels pose significant risks is not addressed. Data from the Montgomery study (11) also must be considered in the context of significant risk.

A Wall Street Journal editoral (14) by Lester Lave, Professor of Economics and Public Policy, Graduate School of Industrial Administration, Carnegie-Mellon University is recommended reading in its entirety. Two principles are proposed: (1) ignore minimal risks, and (2) balance risks against benefits and control costs where risks are not negligible. Professor Lave warns that, "A zero-risk goal will paralyze the entire economy," and refers to safety goals, which also require the careful balance of risk benefit. "Try banning automobiles to end the carnage on the highways. Try banning pharmaceuticals because their side effects pose risks."

The American Chemical Society, in a statement to the EPA Administrator, expressed concern for setting standards without adequate, scientific assessment of risk at low levels of exposure to chemicals in drinking water. "The Society believes that it is not always necessary to develop regulatory controls for potential health hazards that occur at levels where the risks, if any, have not yet been determined."

VOLUNTARY CONSENSUS STANDARDS

William D. Ruckelshaus, in addressing the National Academy of Sciences, noted, "The greatest triumph of a scientist is the crucial experiment that shatters the certainties of the past and opens up rich new pastures of ignorance."

There is need to carefully evaluate - and, where appropriate - regulate and monitor additive contaminants to drinking water. All water treatment chemicals and indirect additive products should be considered in the evaluation process. Decisions must be related to risk, not detection capability. And, there is an established, effective alternative to official regulation through voluntary, consensus standards and objective, third-party testing and certification of related products. The National Sanitation Foundation (NSF) has provided these services through its Standard 14, Plastic Piping System Components and Related Materials (15). This standard establishes minimum public health and safety requirements for thermoset and thermoplastic pipes, fittings, valves, tanks, joining materials, appurtenances, and plastic coatings for potable water and drainage applications. It includes definitions, material and product requirements, and specifies quality control, marking, and record-keeping procedures, as well as testing protocols.

The American Medical Association (AMA), in response to reports from Oregon that persons who complain about disagreeable tastes and odors in drinking water and have suffered apparently related mild diarrheal symptoms may have been affected by contamination from galvanized systems, passed a formal resolution encouraging "national agencies directly involved in standard setting activities such as the National Sanitation Foundation . . . . to develop standards for galvanized steel pipe used for plumbing purposes which would include health aspects as well as physical and chemical properties" (16).

NSF is a private, not-for-profit corporation, chartered in 1944 under Michigan law to develop and administer programs relating to public health and the environment. Through headquarters and laboratories in Ann Arbor, Michigan, a wastewater equipment testing facility in

Chelsea, Michigan, and regional offices in Upland and Davis, California; Ann Arbor, Michigan; Chalfont, Pennsylvania; and Atlanta, Georgia, its programs and services reach throughout the U.S. and into 23 foreign countries. It is best known for its traditional role in voluntary, consensus standards and third-party certification of products conforming with these standards; but, other services consistent with its mission are performed currently, and are planned for future growth. Its organizational integrity and objective services are known and highly regarded around the world.

A highly qualified professional and technical staff - engineers, chemists, microbiologists, and environmental scientists - is retained by NSF, and state-of-the-art facilities are retained or available. Faculty members from The University of Michigan are retained for consultant toxicological, radiological, and other expertise, and the University facilities are available as required. Recent capital acquistions focus on laboratory and information processing resources intended to optimize quality and timeliness of all operations.

Plastics piping systems have been tested by NSF since 1955 - first in a comprehensive special study which included animal feeding of water exposed to plastics (17), then, under NSF Standard 14 (15), for which the basic toxicological and leachate data were provided by the special study. An annual product Listing of items tested under Standard 14 and authorized to bear the NSF logo is published (18) and widely circulated. It is not uncommon for code and regulatory authorities to reference or require conformance with an applicable NSF standard, or to require that products appear in the current Listing. Standard 14 was adopted in 1965 and is revised on a continuing basis.

In the plastics program, there is an understanding between NSF and the EPA Office of Drinking Water, that contaminants regulated by the National Primary Drinking Water Regulations (NPDWR) and associated with plastics products listed by NSF under Standard 14 for potable water applications, will be monitored by leachate or direct chemical testing, and meet the MCLs specified in the NPDWR. The procedure followed by NSF in carrying out this agreement has been to closely monitor the NPDWR; and, when new MCLs are promulgated, determine by formulation review and special testing whether or not they are associated with any types of products listed for potable water end use.

Under NSF Standard 14, ingredients proposed for use in potable water formulations are accepted by comprehensive review and testing protocols, then qualified for use in listed products. These requirements are summarized in Figure 1. With an application for listing, the manufacturer provides complete, detailed chemical identity of all ingredients in the formulation. This information is reviewed by staff, and held under terms of confidential

Acceptance[1] *(New or Generically Similar Ingredient)*	Qualification[1] *(New Product or Change in Formulation)*	Monitoring[2]
Disclosure statement with complete chemical identity		
Product and compound tested at 2 X maximum recommended use level	Product and compound tested at maximum use level	
Chemical leachate testing[3] 1st exposure ≤10 X MPL[4] 3rd exposure ≤MPL[4]	Chemical leachate testing[3] 1st exposure ≤10 X MPL[4] 3rd exposure ≤MPL[4] Performance testing HDS ASTM[4] ASTM[4]	Chemical extraction testing[3] 3rd exposure ≤MPL[4] Performance testing HDS ASTM[4]
FDA GRASd If no, Ames test and 90-day feeding study at effect, no effect, and intermediate levels		
[1]Samples submitted by applicant. [2]Samples collected by regional staff. [3]RVCM ≤MPL is also required for PVC and CPVC pipe and fittings. [4]Refers to all parameters in Standard 14.		

Figure 1. Summary of requirements for products listed under Standard 14.

disclosure. For ingredients generally recognized as safe (by FDA) for food contact, (GRAS), no further toxicological testing usually is required. For non-GRASd ingredients, Ames test data (for mutagenicity) and animal feeding studies must be provided. The animal data must include established no-effect, effect, and intermediate levels of the proposed ingredient; the protocol and testing laboratory must be approved in advance by NSF's retained toxicologists.

Chemical leachate testing is required for all proposed new ingredients. Pipe and compound are formulated to maximum use. [For example, if a stabilizer supplier proposes a new product for use in pipe recipes at a level ranging from 2 to 4 parts per hundred parts of resin (phr), he must submit samples of pipe and compound containing the stabilizer at 8 phr.] Formal certification statements are required and are retained on file for both maximum use and two-times use levels in the products and materials tested.

Chemical leachate (extraction) testing is performed consistent with the testing protocol in NSF Standard 14, i.e., exposure to "formulated water" at pH 5.0 ± 0.2, at 37 ± 0.5 degrees Centigrade (°C), for periods of 24, 24, and 72 hours.[a] Fresh water is added following each of the 24-hour

---

[a]Assumes cold water end use applications. For hot water applications exposures are 1 hour at 82°C, and 0.5 hour at 82°C followed by 72 hours at 37°C.

exposures, providing data for three separate extractions. The first exposure is assumed to simulate worst case end use, where the user would ingest the first, very aggressive water drawn from a new installation; the third exposure represents water which could be ingested as the first draw of water left standing in pipes over a weekend.

The chemical parameters monitored and their maximum permissable limits (MPLs) are shown in Table II. For proposed new ingredients, the MPL for the first exposure (MPL-1) is set at ten times the established, third-exposure MPL (MPL-3). MPL-3 is equivalent to maximum contaminant levels (MCLs) in the NPDWR for all regulated chemicals included in NSF Standard 14.

The addition of total trihalomethanes (TTHMs) to NSF Standard 14 is a "good faith" modification to the policy of adding regulated chemicals (MCLs) established in the NPDWR to the Standard if the contaminants are known to be associated with listed products.

TABLE II. CHEMICAL PARAMETERS IN STANDARD 14

PARAMETER	MPL (ppm)	
	-1	-3
Antimony	0.5	0.05
Arsenic	0.5	0.05
Barium	10.0	1.0
Cadmium	0.1	0.01
Chromium	0.5	0.05
Lead	0.5	0.05
Mercury	0.02	0.002
Phenolic Substances	0.5	0.05
RVCM*	10.0	10.0
Selenium	0.1	0.01
Tin	0.5	0.05
TTHM	1.0	0.10

*In finished product

To test for volatile organic chemicals, a valid exposure protocol was required. An NSF report, Proposed Organohalide Leaching Test Protocol for Plastic Piping (10) describes a capped pipe procedure and validation data for 19 pipe samples - 7 PVC and 12 CPVC. A jar test technique currently is being validated. All data acquired to date indicated that TTHMs are not leached by plastic pipe at significant levels; but, because TTHMs are regulated, and because there is considerable current concern for organic chemical contaminants in drinking water, an MPL was recommended for NSF Standard 14 to assure collection of an adequate representative data base for plastic pipe.

The procedure for accepting proposed new ingredients is diagrammed in Figure 2. To confirm compatability with other ingredients in a product recipe, an accepted new ingredient then must be "qualified" by additional chemical leachate testing. Products formulated by the ingredient user (pipe, fittings, or appurtenance manufacturer) at the maximum proposed use level are tested in accordance with the procedure described for acceptance testing. (This procedure is shown in Figure 3.)

Following acceptance of an ingredient and qualification of product, listing is authorized. The appropriate logo is then displayed on the product,[a] a minimum of three annual, unannounced inspections are made at each production location, and samples are selected, indelibly marked in the production plant by the NSF representative at the time of selection, and shipped to Ann Arbor for complete chemical and physical testing in accordance with NSF Standard 14. The inspection and sampling functions are handled by qualified personnel from the respective regional offices.

To demonstrate whether or not the organic chemicals proposed for regulation by EPA are leached from potable water products listed by NSF, an additional monitoring requirement has been added to Standard 14 for the five VOCs recommended for regulation by the National Drinking Water Advisory Council (NDWAC) (20). This requirement is for data acquistion only. The Council's position is that both health effects and occurrence data should be available to support the setting of standards. Therefore, of the 15 VOCs included in the March 1982 ANPRM, only carbon tetrachloride (CCl_4) trichloroethylene (TCE), tetrachloroethylene (PCE), 1,1,1-trichloroethane, and 1,2-dichloroethane were recommended by NDWAC for regulation. Neither RMCLs nor MCLs have been propsed or promulgated, and MPLs have not been added to NSF Standard 14.

---

[a] NSF - pw, - wc, for potable water/well casing applications; NSF - dwv. - tubular, for drain, waste, and vent/continuous waste systems. NSF-cw for corrosive wastes; NSF-sewer for sewer main applications.

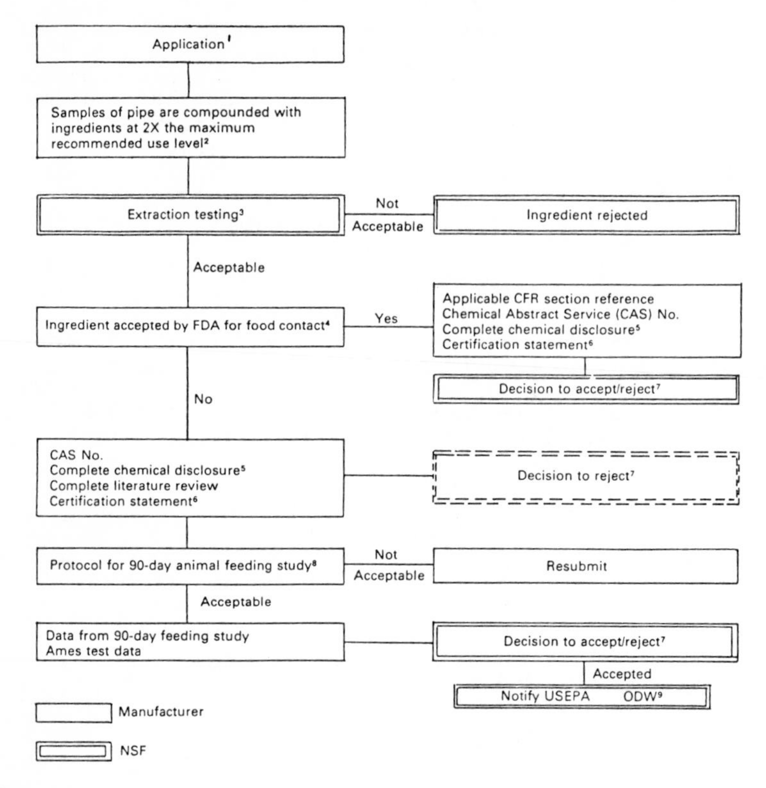

Notes:

1. Potable water applications only.
2. Manufacturer must provide certification statement with level of ingredient in samples and maximum recommended use levels.
3. Extraction testing protocol in Standard 14; decision criteria include ingredient(s) extracted ≤ 10 x MPL in Standard 14 in 1st exposure and ≤ MPL in 3rd exposure.
4. US Code of Federal Regulations (CFR), Title 21 (Food and Drug Regulations).
5. Includes chemical description and levels with molecular structure; purity; identification and levels of all known contaminants; identification of known potential carcinogens with appropriate literature references.
6. Manufacturer must provide certification statement that proposed ingredient is suitable for potable water end use applications.
7. Additional analytical and/or toxicological data, risk benefit assessments, and health effects information may be required.
8. Three levels of proposed ingredient are required — effect, no effect, and an intermediate. Feeding study may not be required for acceptance of generically similar ingredient.
9. Office of Drinking Water (ODW).

Figure 2. Procedure for acceptance of proposed new and generically similar ingredients -- potable water applications only.

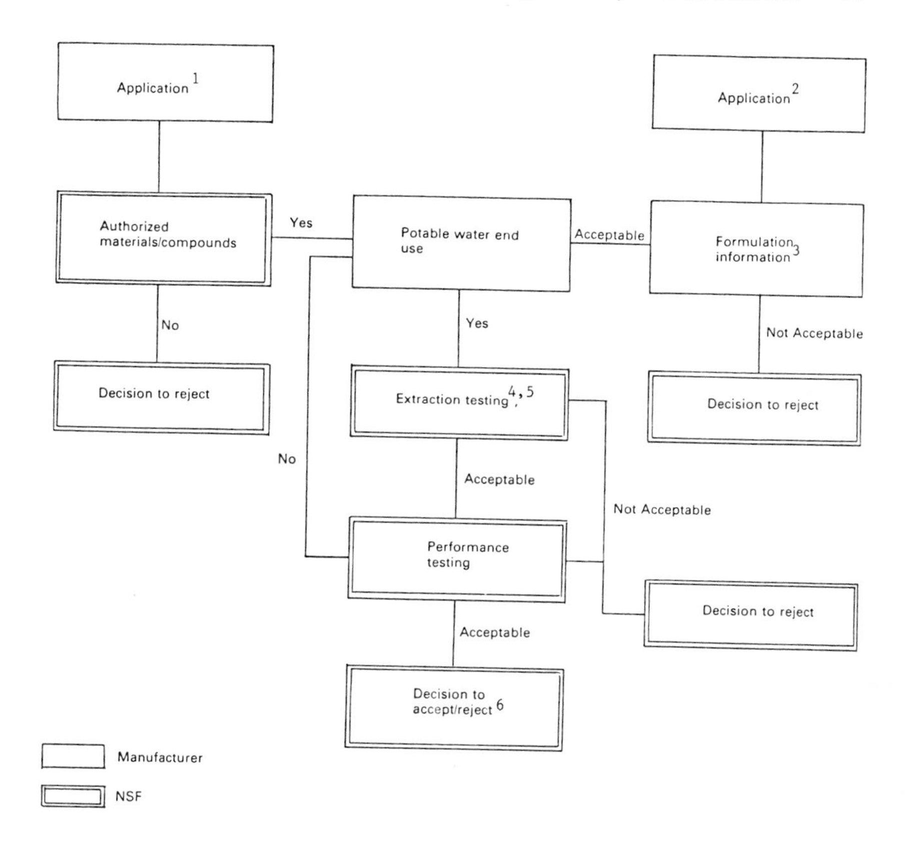

Notes:

1. Thermoplastic pipe and fittings.
2. Thermoplastic materials, compounds, solvent cements, thermoset pipe and fittings, ingredient changes.
3. Formulation information is submitted on the form, "Confidential Report on Plastics Formulation for Materials"; or on the form, "Confidential Report on Plastics Formulation for Finished Product".
4. Manufacturer must certify the formulation and use levels of ingredients in the submitted samples.
5. Extraction testing protocol in Standard 14; decision criteria include ingredient(s) extracted ≤ 10 x MPL in Standard 14 in 1st exposure and ≤ MPL in 3rd exposure.
6. Materials or compounds for pressure piping applications must also have an established hydrostatic design basis (HDB).

Figure 3. Procedure for qualification of materials, products and alternate ingredients -- for potable water and drainage application only.

In 1978, a requirement for residual vinyl chloride monomer (RVCM) was added to NSF Standard 14. By modeling and leachate testing, it was established that levels to 10 parts per million (ppm) RVCM in the wall of pipe or fittings would not leach detectable levels of the monomer to water exposed to product (where "detection" is 2 ppb). Results of RVCM monitoring experience are shown in Table III. It is significant that listed products have decreased significantly - from occasionally greater than 100 ppm RVCM (1977-79) to consistently less than 10 ppm - over the period of NSF Standard 14 testing. No samples tested in 1982 and 1983 to date have exceeded the established MPL.

The types of plastics commonly accepted for potable water end use - PVC, CPVC, PE, and PB - and typicial formulation ingredients for each are summarized in Table IV. Heat stabilizers are essential ingredients in PVC and CPVC pipe, fittings, and appurtenances. They are typically organometallic compounds of tin or antimony, such as : (mono- and/or di-) -methyltin, -butyltin, or -octyltin esters or estertin, compounds. There are also inorganic formulations such as calcium-zinc which are in liquid suspension with hydrocarbon diluents or a mixture of barium carbonate-barium alkylphenolate. Lead-based stabilizers, although commonly used outside the U.S., have not been accepted by NSF. Lubricants in PVC may be calcium stearate, paraffin wax, or EP wax; in ABS, they are Acrowax-C, or magnesium stearate. The PVC filler typically is calcium carbonate.

TABLE III. RVCM EXPERIENCE DATA

PERIOD	NUMBER SAMPLES TESTED (n)	NUMBER SAMPLES (ppm)				
		≤10	>10<20	>20<50	>50<100	≥100
1977	200	182	2	5	6	5
1979	650	629	11	7	1	2
1980	457	455	1[a]	1[b]	0	0
1981	474	473	1[c]	0	0	0
1982	285	285	0	0	0	0
1983*	150	150	0	0	0	0

[a]14.4 ppm; retest sample <10

[b]27.1 ppm; retest sample <10

[c]15.7 ppm; retest sample <10

*January through July only

TABLE IV. INGREDIENTS IN PW PLASTICS

Type	Ingredient								
	Resin	Stabilizer	Antioxidant	Lubricant	Filler	Processing Aid/ Impact Modifier	Pigment	Tracer	Solvent
PVC	X	X		X	X	X	X		
CPVC	X	X		X	X	X	X		
PB	X		X			X	X	X	
PE	X		X			X	X	X	
ABS	X		X	X			X		
Cement/ Primer	X						X		X

The processing aids or impact modifiers in PVC are polymers, such as chlorinated polyethylene (CPE), ABS, methyl methacrylate-butadiene-styrene, or alpha-methylstyrene. Pigments are typically titanium dioxide or carbon black; but others are accepted, including, for example, talc (hydrous magnesium silicate specified to contain zero fibrous content), ferric oxide, or copper thiocyanine in polybutylene.

Antioxidants are required in the polyolefins (PE and PB) and ABS. In the polyolefins, they are hindered phenols, such as Irganox 1010 [tetrakis-(methylene-3-{3', 5'-ditert.-butyl-4-hydroxyphenyl}propionate)methane]. In ABS, they are typically dilauryl thiodipropionate.

The polyolefin materials and compounds in PW products are required by NSF to contain innocuous trace elements. These are inorganic metals or metallic oxides added at assigned levels of 10 to 1,000 mg/L. The tracer and level are specific to each manufacturer (material supplier or in-plant compounder), and are included in the registered formulation for each material or compound. Products (pipe and fittings) manufactured from these materials or compounds are analyzed for tracer levels to verify the source of material or compound used. In-plant compounds that are blends of listed materials also are analyzed to verify the source and approximate levels of the source materials. The assigned tracers include zirconium oxide, aluminum, tin oxide, magnesium carbonate, zinc oxide, vanadium, molybdenum, barium carbonate, and titanium dioxide.

The inorganic chemicals in the NPDWR and others peculiar to the PW listed plastics (e.g., tin and antimony used as heat stabilizers) are measured routinely in leachate testing (acceptance, qualification, and monitoring). Both leachate and performance testing are required for PW products; performance only is required for drainage products. Failure experience for listed samples - 1980 through July 1983 - is shown in Table V.

TABLE V. SUMMARY OF RECENT EXPERIENCE FOR CHEMICAL LEACHATE TESTING UNDER NSF STANDARD 14

Period	Samples Tested (n)	Samples Failed (nF)	Failed Samples (%)	Parameters Failed		
				Tin (Sn) (nF)	Cadmium (Cd) (nF)	Phenols (nF)
1980	491	9	1.8	9	0	0
1981	694	3	0.4	3	0	0
1982	407	8	2.0	6	1	1
1983*	229	2	0.9	1	0	1

*January through July only

The three annual inspections of production facilities are minima, and apply to foreign and domestic sites; nonconformance and other problems result in additional inspections. At the plant, products in production and inventory are checked to verify consistency with previous design and testing records. Use of only accepted ingredients or compounds, and in-plant quality assurance are verified; and selected evaluations may be performed (e.g., dimensioning plastics). Plastics plant inspections in 1982 numbered 711; 736 are planned for 1983. Violations identified either in testing or inspection result in various actions, ranging from corrective measures to delisting.

Solvent systems - cements and primers - are used to join PVC, CPVC, and ABS products. Ingredients in these products include the solvents (principally THF, cyclohexanone, MEK, or DMF), resin or compound, and pigments (principally titanium dioxide or carbon black). By policy, label verification to qualify and quantify the solvents will be required monitoring practice under NSF Standard 14.

Recognition of the desirability and feasibility of uniform, minimum national standards for products and services relevant to health of the public and quality of the environment is a principal corporate tenet. To achieve success, understanding and appreciation of the needs of all parties concerned is essential. From the outset, standards development at NSF has included representatives at all levels of government, the affected industry, and users of the subject products or services. Typically, the request for a standard may originate with any of the three sectors; interest and commitment are established in an exploratory meeting. A small task committee then is appointed to draft the standard. The draft is reviewed and revised or accepted by the Joint/Industry Advisory Committees (JC/IAC), where each sector - regulatory, industry, and user - has voting representation. The Joint Committee proposes the standard to the Council of Public Health Consultants, a group of 28 public health professionals with no industry representation. The Council recommends a standard to the Board of Trustees, where final adoption occurs. This process is diagrammed in Figure 4.

New standards development activities are classified by anticipated complexity. Completion of the most complex is targeted for two years; the least complex, one year. A modified Project Evaluation Review Technique (PERT) has been adopted by the program staff as a means of achieving the completion targets.

All NSF standards have a requirement for periodic review at intervals not to exceed five years. The formal, on-going review and revision - or reaffirmation - process is accomplished through established Industry Advisory (IAC) and Joint Committee procedures. All voting is in compliance with procedures described in the Office of Management and Budget (OMB) Circular A-119.

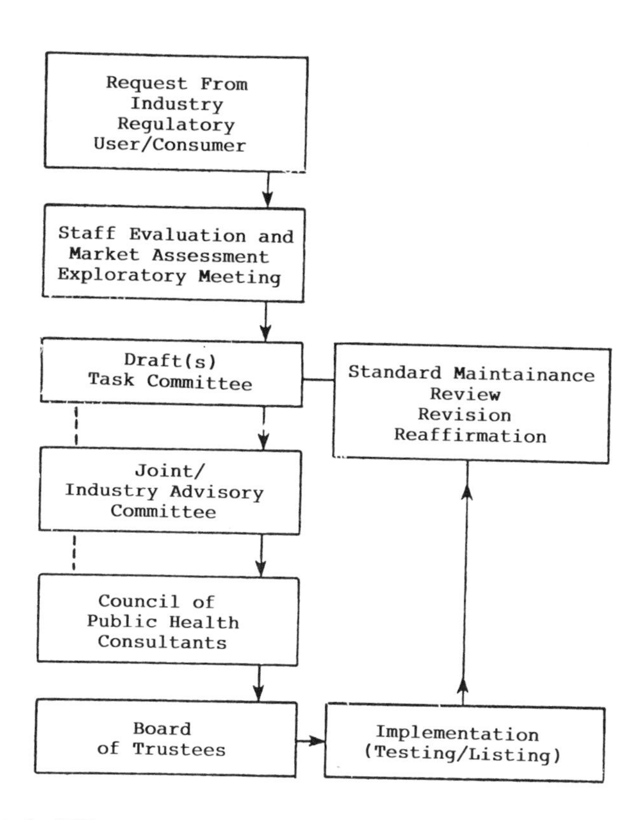

Figure 4. NSF consensus standards process

The type of comprehensive, cooperative regulatory - third-party - industry program in place for plastics piping system components for more than 20 years illustrates an "innovative regulatory reform" consistent with the current Washington rhetoric. It effectively and appropriately places cost burdens with the private sector; and, credibility is assured by the third-party participant. It is, "an effective alternative to official regulation of indirect additives to drinking water."

REFERENCES

1. 44 CFR 141, pp.427775-8, July 20, 1979.
2. Advance Notice of Proposed Rulemaking, 40 CFR 141, Federal Register 47(43):9350-9358 (March 4, 1982).
3. Committee on Water Treatment Chemicals, Water Chemicals Codex, (Washington, DC; National Academy Press, 1982).
4. Rehwoldt, R., "Water Chemicals Codex", presented at the 168th National Meeting of the American Chemical Society, Washington, DC, August 1983.
5. Nicholas, D., "Corrosion of Copper Water Tubes in the Hunter Region of NSW", presented at the annual meeting of the Australia Corrosion Control Association, November 1980.
6. National Sanitation Foundation, "Preliminary Study on Comparative Leaching Characterisitics of Plastics, Copper, and Galvanized Steel Piping System Components", September 1980.
7. Wong, P. and C.S. Berrang, "Contamination of Tap Water by Lead Pipe and Solder", (journal and date not available).
8. Herrera, C.E., et al, "Seattle Distribution System Corrosion Control Study: Volume III. Potential for Drinking Water Contamination from Tin/-Antimony Solder", EPA Report No. 600/S2-82-018, August 1982.
9. Herrera, C.E., et al, "Evaluating the Potential for Contaminating Drinking Water for the Corrosion of Tin-Antimony Solder", J. Am. Water Works Assoc. 74(7):368-375 (1982).
10. Papan, L., "Assembly Concurrent Resolution No. 98", California Legislature 1979-80 Regular Session, January 15, 1980.
11. James M. Montgomery Consulting Engineers, Inc. Pasadena, CA, "Solvent Leaching from Potable Water Plastic Pipes", (October 1980).
12. Brown, S.L., "Environmental Review of Proposed Expanded Uses of Plastic Plumbing Pipe", (Menlo Park, CA; SRI International, March 1983).

13. Boettner, E.A., G.L. Ball, et al, "Organic and Organotin Compounds Leached from PVC and CPVC Pipe". EPA Report No. 600/51-81062, February 1982.
14. Lave, L., "The High Cost of Regulatory Low Risks", The Wall Street Journal, Friday, August 19, 1983 (p. 14).
15. National Sanitation Foundation, Standard Number 14, Plastic Piping System Components and Related Materials, revised August 7, 1983.
16. Sammons, J. H., MD, American Medical Association, personal communication, May 29, 1980.
17. National Sanitation Foundation, "A Study of Plastic Pipe for Potable Water Supplies", June 1955.
18. National Sanitation Foundation, "Plastic Piping System Components and Related Materials, 1982 Listing."
19. National Sanitation Foundation, "Proposed Organohalide Leaching Test Protocol for Plastic Piping", April 1983.
20. National Drinking Water Advisory Committee, Summary and Minutes of Meeting, September 23-24, 1982. (Washington, DC; U.S. EPA Office of Drinking Water).

CHAPTER 7

THE WATER TREATMENT CHEMICALS CODEX

Robert E. Rehwoldt, P.E.

Food and Nutrition Board
Commission on Life Sciences
National Academy of Sciences
Washington, DC 20418

The availability and production of potable water are matters of great national and worldwide concern. In the U.S. alone, an estimated 1-2 billion gallons of drinking water must be provided each day. To comply with health and other applicable standards for treating that amount of potable water, suppliers used more than 1.2 million tons of chemicals in 1981.

In 1980, the National Academy of Sciences Committee on Water Treatment Chemicals was formed and entrusted with the task of developing specifications, first for direct additives, and later, as feasible, for indirect additives. After two years of deliberations, the committee produced a Water Treatment Chemicals Codex. The Codex is meant to supplement existing compendia on water treatment chemicals and is confined to information on purity as it is related to health. It does not address product performance, packaging, storage, or handling.

To arrive at its recommended contaminant limits, the committee met with the U.S. EPA with the aim of compiling a list of priority chemicals. This list then was categorized according to use pattern, that is, those chemicals used in coagulation and flocculation; softening, precipitation and pH control; disinfection and oxidation; and miscellaneous treatment applications. In drafting the monographs in each category, a subgroup of the committee reviewed current data on known impurities in the chemicals, grades of manufactured products, use patterns, and other variables.

The committee also developed a list of impurities to be considered. Initially, the list was identical to that of the regulated inorganic impurities specified by the National Interim Drinking Water Regulations developed in response to the Safe Drinking Water Act of 1974. This list subsequently was modified to include those substances for which there is evidence of occurrence as contaminants in water treatment chemicals. The toxicology subgroup of the committee supplied toxicological data on these substances, including information on possible genotoxic or epigenetic (nongenetic cellular damage) effects.

In general, the committee felt that it would be appropriate to utilize the Maximum Contaminant Level (MCL) for calculating the allowable contaminant level contributed by an impurity in a water treatment chemical, unless there was no current MCL for that impurity, or when there was new information concerning either the toxicity of the contaminant or the current status of the MCL.

An MCL thus was converted to a Recommended Maximum Impurity Content (RMIC) for the additive by the following equation:

$$\text{RMIC} = \frac{\text{MCL}}{\text{maximum dosage x safety factor}}$$

$$= \frac{\text{MCL (mg/L) x } 10^6 \text{ mg/kg}}{\text{MD (mg/L) x SF}}$$

The maximum dosage for the water treatment chemical was based on maximum patterns known by the committee to be representative of water treatment practice. The safety factor used in the calculation of the RMIC was 10, reflecting the view of the committee that no more than 10% of a given MCL value should be contributed by a given impurity in a water treatment chemical.

The Codex contains RMIC values for impurities of concern at selected additive dose levels, which are reported to one significant figure. RMIC values defining the purity of each water treatment chemical also are contained in individual monographs and may be used as guidelines for the water works industry. At present, the Codex contains 34 monographs; an additional 8 are expected to be completed by the end of the year.

Analytical procedures were selected from compendia on methodology and protocol, adopted from manufacturers or derived from methods set forth in the scientific literature. Data on toxicological aspects were obtained from the scientific literature, from chemical manufacturers, and from the Code of Federal Regulations.

The RMIC levels are based upon information available to the committee. It is impossible to recommend maximum content levels for unusual or unexpected impurities, the presence of which would depend upon the method of manufacture and the quality of raw materials used. If unusual raw materials or unusual methods of manufacture go into the preparation of a treatment chemical, the user should require appropriate certification of purity from the vendor or manufacturer, to prove that the chemical is suitable for application to the making of potable water.

It is expected that the Codex will be reviewed continuously and that annual supplements will be issued. The supplements may contain lists of additional chemicals and revisions of the monographs contained in the present

Codex, as well as revisions of analytical procedures. Information about ordering it can be obtained from the National Academy of Sciences, Publications Office, Washington, DC 20418.

CHAPTER 8

DISINFECTANT CHEMISTRY IN DRINKING WATER-OVERVIEW OF IMPACTS ON DRINKING WATER QUALITY

A.A. Stevens, L. Moore, R.C. Dressman, and D.R. Seeger

U.S. Environmental Protection Agency
Municipal Environmental Research Laboratory
Drinking Water Research Division
Cincinnati, Ohio 45268

Chemicals commonly considered for use as disinfectants in municipal drinking water treatment are chlorine, chloramines, chlorine dioxide, and ozone. Considerations such as disinfection power, ease of application, and low cost in the past have led to the use of free chlorine ($HOCl/OCl^-$) as the primary disinfectant. Discovery of trihalomethanes, formed by the action of free chlorine upon natural organic materials such as aquatic humic materials, has led to a reexamination of this practice. In many cases a change to an alternative disinfection practice either has occurred or is being contemplated by utilities that otherwise have difficulty meeting the maximum contaminant level requirement for trihalomethanes. Chloramines, chlorine dioxide, or ozone when used alone do not usually cause significant formation of trihalomethanes. Each is an active oxidant, however, and has the potential of forming objectionable byproducts other than trihalomethanes (as may chlorine as well). Free chlorine, for example, forms quantities of "organic halogen" substantially in excess of those accounted for by the trihalomethanes. Chloramines and chlorine dioxide do also, although to a lesser extent than free chlorine. Specific reaction products of each in aqueous solution now are being identified. Thus, the beneficial use of each of these chemicals as a biocide in drinking water treatment may have its own less desirable side effects.

INTRODUCTION

When we were first asked to present a paper at this symposium on the subject of disinfectant chemistry in drinking water as an "overview" paper, we had in mind a somewhat traditional textbook approach describing in a brief way all the known basic chemistry. After reminding ourselves of the complexity of that, including the problem of distilling the huge volume of literature on that subject, we decided to take a different approach that we think is more appropriate for this group and better addresses the thrust of this symposium. That is, we plan to address the various perceived impacts of the application of disinfectant chemicals on drinking water quality in a comparative way and, to the extent it is understood, to relate these observations to the known chemistry of each disinfectant. The basic chemistry of the oxidants in water treatment, their uses as oxidants, as disinfectants, and of trihalomethane formation and control have been reviewed in a much more complete and detailed fashion elsewhere (1-5), and this material served as the source of much of the following discussion.

DISINFECTANTS

Four chemical oxidants are considered most often as drinking water disinfectants. These are chlorine (free), chloramines ("combined chlorine"), chlorine dioxide, and ozone.

Free Chlorine

Free chlorine exists in aqueous solution as a mixture of hypochlorous acid (HOCl) and hypochlorite (OCl^-) ion:

$$Cl_2 + H_2O \longrightarrow HOCl + H^+ + Cl^- \qquad [1]$$

$$HOCl \rightleftharpoons H^+ + OCl^- \qquad [2]$$

The source can be chlorine or a hypochlorite preparation such as NaOCl (solution) or $Ca(OCl)_2$. In either case, the final pH of the drinking water establishes the relative amounts of each of the active species. At pH 7.5 the two exist in almost equal concentrations.

Chloramines

The chemistry of "combined chlorine" is essentially the combined chemistries of free chlorine and ammonia. The ammonia may be present naturally in the source water or added intentionally by the utility.

Free chlorine first reacts with ammonia (NH_3) to form monochloramine (NH_2Cl) by the substitution reation:

$$HOCl + NH_3 \rightleftharpoons NH_2Cl + H_2O \qquad [3]$$

Sequential reactions occur to form dichloramine ($NHCl_2$) and nitrogen trichloride (NCl_3). Oxidation occurs in the process when sufficient excess chlorine is present to remove the ammonia as nitrogen gas:

$$2NHCl_2 + H_2O \longrightarrow N_2 + HOCl + 3H^+ + 3Cl^- \quad [4]$$

All of this is somewhat simplified, and the end result is heavily dependent on pH. Nevertheless, for the purposes of this paper it is sufficient to understand that free chlorine exists only in very small quantity (relative to that added) until the ammonia is decomposed and removed as nitrogen gas by the process well known in the drinking water industry as breakpoint chlorination (Figure 1).

Chlorine Dioxide

Chlorine dioxide (ClO_2) is a gas at normal temperature and pressure and exists as molecular ClO_2 in pure aqueous solution. Pure chlorine dioxide is unstable and can be an explosive mixture in air at approximately eleven percent concentration (4). For that reason it is produced on-site at a utility. This usually is by reaction of sodium chlorite with excess chlorine in acid solution:

$$HOCl + H^+ + 2ClO_2^- \longrightarrow 2ClO_2 + H_2O + Cl^- \quad [5]$$

Although disproportionation of ClO_2 to form ClO_2^- and ClO_3^- takes place, this reaction is much slower and, therefore, is less important from a chemical disinfectant standpoint than the nearly immediate hydrolysis of molecular chlorine to hypochlorite. Chlorine dioxide does not react with ammonia, but ammonia-free waters do exhibit a demand for ClO_2 similar in magnitude to that for free chlorine (3).

Ozone

Molecular ozone (O_3) is a very strong oxidant and relative to the above disinfectants has a very short half-life in water - said to be on the order of twenty minutes at low pH values. Molecular ozone is the reactive species at neutral and lower pH values, reacting through an ionic mechanism. At pH values above 8-9, O_3 reacts with water to produce hydroxyl and hydroperoxide radicals as the active species (4).

RELATIVE DISINFECTION EFFECTIVENESS

The primary reason for the use of disinfectants in the treatment of drinking water is to ensure the destruction of pathogenic microorganisms during the treatment process, thereby preventing the transmission of disease by drinking water. Secondarily, the presence of a disinfectant in the water distribution system helps to maintain the quality of water by preventing the growth of nuisance microorganisms.

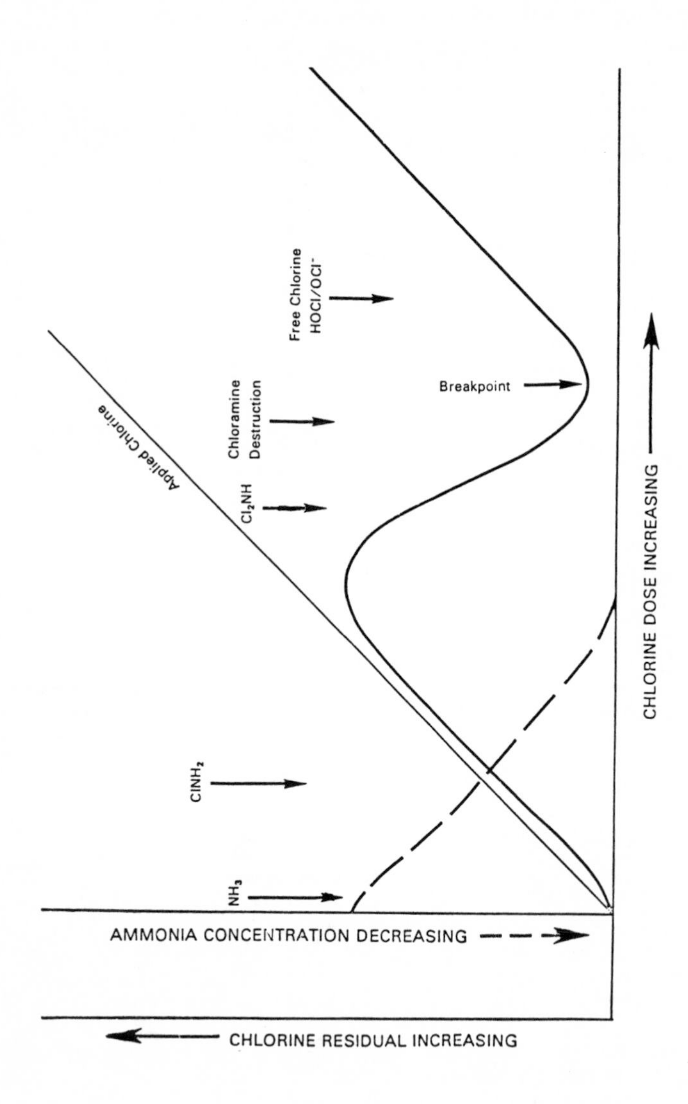

Figure 1. Chlorine - ammonia breakpoint curve.

The effectiveness or efficiency of biocidal agents is determined by the rate at which the reaction or killing of the microorganism population proceeds (3). The comparative biocidal efficiencies of disinfectants frequently are expressed as the relative concentration (mg/L) of different disinfectants needed to obtain equivalent disinfection rates, or as the relative inactivation rates produced by the same concentration of different disinfecting agents. Most of this information has been obtained by laboratory experimentation under carefully controlled conditions, which include clean systems, the absence of extraneous disinfectant-demanding substances, and the use of pure cultures of the microorganism under study. The presence (in solution) of materials exerting disinfectant demand is likely to change disinfection efficiencies by way of competing reaction mechanisms. This effect complicates extrapolations from experiments with clean systems to expected water utility performance. Nevertheless, comparisons of disinfectant performance under laboratory conditions are instructive.

Data from results of a number of experiments conducted using different disinfectants at various concentrations can be used to construct plots of the type shown in Figure 2. These results show the exposure times and concentrations of several disinfectants needed to produce a given level of inactivation of a given microorganism. The results show that chlorine dioxide at pH 7 and HOCl at pH 6 produce similar rates of inactivation of *Escherichia coli*. Hypochlorite ion (OCl^-) at pH 10 was less effective, and monochloramine at pH 7 and dichloramine at pH 4.5 were even less so. A similar plot showing virucidal efficiency of these disinfectants for poliovirus 1 is shown in Figure 3. Note that in general, higher disinfectant concentrations and longer contact time are needed for inactivation of poliovirus 1 than for *E. coli*.

A comparison of the relative positions of the curves in Figures 2 and 3 reveals that the relative efficiencies of the various disinfectants vary according to the target organism. Ozone is not included in these figures because fast reaction rates and disinfectant instability cause limitations on sampling times and ozone measurements, thus making good experimental results difficult to obtain. Ozone does, however, inactivate these microorganisms at a high rate.

Assessing the efficiencies of different free and combined chlorine species also is complicated by the nature of the chemical reactions that determine the chemical species present and the chemical equilibria established under various pH conditions. For instance, during the reaction of equation [2] a rapidly achieved equilibrium exists that is drastically influenced by pH. At pH 10, approximately 0.5 percent of the free residual chlorine still is present as HOCl, and because it is a much more powerful biocide than OCl^-, its presence could influence the observed biocidal activity substantially.

Similarly, equation [3] is reversible, and a solution of 2 mg/L NH_2Cl is estimated to be 0.58 percent hydrolyzed (0.58 percent HOCl) at pH 7 and 25° C. Because of the much higher biocidal efficiency of HOCl, its influence on the disinfection rate observed could be substantial and could explain the influence of pH on the biocidal efficiency of monochloramine.

Futhermore, equation [6] indicates that although

$$H^+ + 2NH_2Cl \rightleftharpoons NH_4^+ + NHCl_2 \qquad [6]$$

mostly monochloramine is formed when excess ammonia is present at high pH (>8), addition of hydrogen ion (lowering pH) will cause formation of dichloramine, with the position of this equilibrium being determined by the pH of the

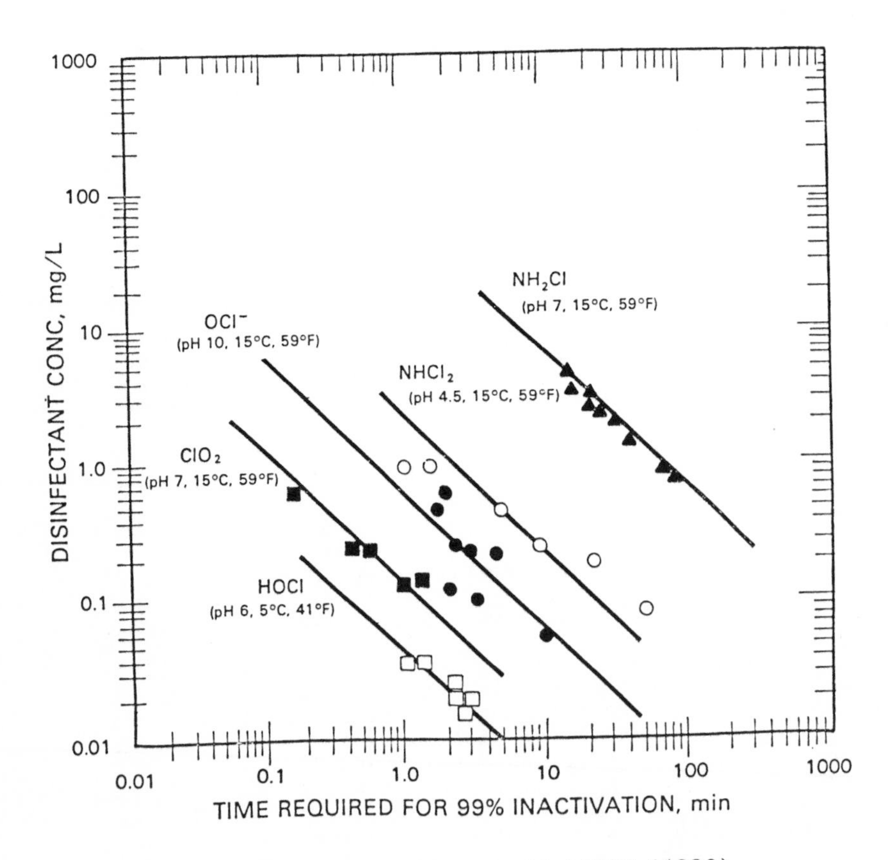

Figure 2. Inactivation of E. coli (ATCC 11229) (from ref. 3, based on refs. 6, 7).

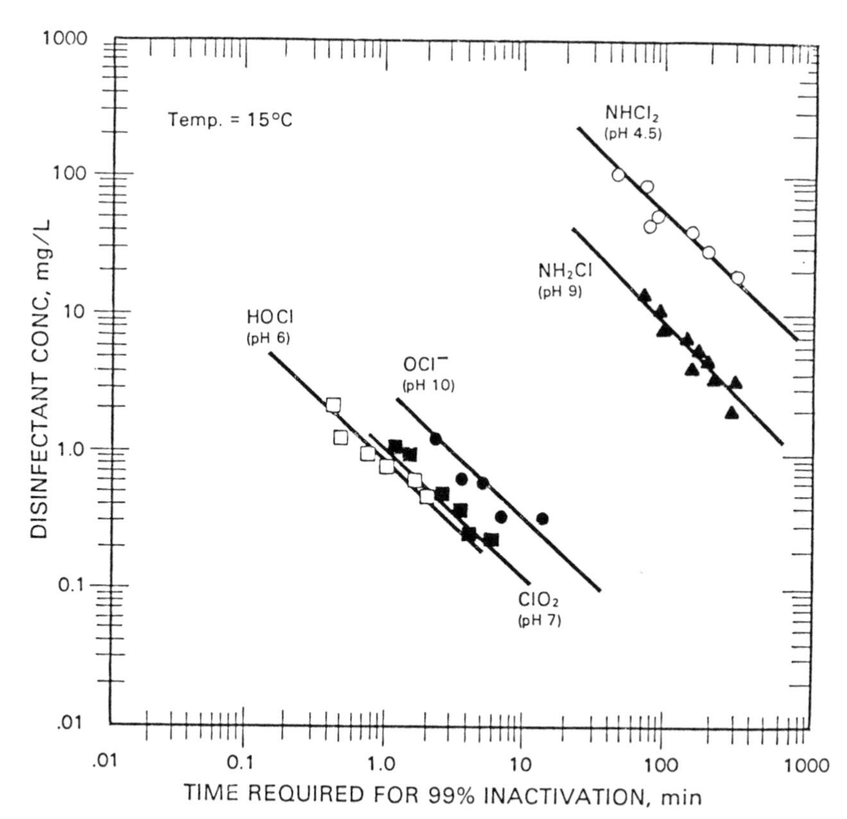

Figure 3. Inactivation of poliovirus 1 (Mahoney) (from ref. 3, based on refs. 6. 7).

treated water. Thus with chlorine and chloramines, pure species never are present, and pH determines which are present and their proportions. The influence of pH, therefore, cannot be experimentally separated from species effectiveness for disinfection.

Nevertheless, in the case of chlorine, disinfection efficiency declines rapidly as the pH is increased from 7 to 9. The efficiency of chlorine dioxide also changes substantially over this pH range; but in contrast to chlorine, the effectiveness increases as the pH increases (Figure 4). In this case, the change appears to be in microorganism sensitivity rather than in disinfectant species present because, unlike chlorine, chlorine dioxide does not so rapidly dissociate or disproportionate into different chemical species within this pH range.

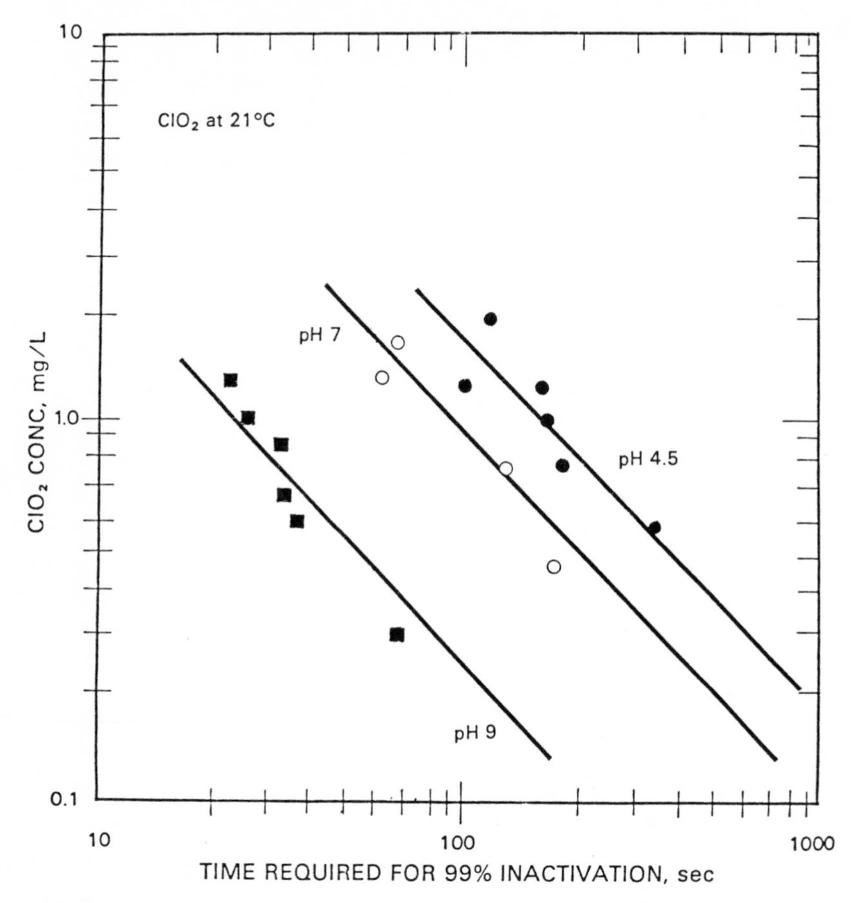

Figure 4. Effect of pH on inactivation of poliovirus 1 (Mahoney) with chlorine dioxide (from ref. 3, based on ref. 6).

The pH of the water also affects ozone chemistry. At high pH values, ozone decay is accelerated, proceeding through radical ion intermediates; thus, the pH of the water being treated may also influence ozone effectiveness.

OTHER BENEFICIAL USES

Although the primary purpose of adding these oxidants to drinking water during treatment is for the control of microbiological contaminants, especially pathogenic organisms, certain side benefits also accrue (1-5). Source waters high in ferrous iron and manganese are treated successfully with ClO_2, O_3, and free chlorine by oxidative precipitation. Ozone has been observed to improve the clarification process. Chlorine can remove ammonia, and hydraulics are improved or maintained by control of slimes and macroorganisms. Ozone and chlorine dioxide are used to remove or prevent taste and odor. In some locations the oxidants are used to remove color, and it is this latter case that hints of the occurrence of some less desirable side effects.

NEGATIVE ASPECTS OF DISINFECTANT USE

Many considerations, such as disinfection power, ease of application, and low cost, in the past have led to the heavy use of free chlorine ($HOCl/OCl^-$) as the primary disinfectant in drinking water treatment, often very early in the treatment process. Discovery of trihalomethanes, however, formed by the action of free chlorine upon natural organic materials, including aquatic humic materials that are responsible for undesirable color in many source waters, has led to a reexamination of this practice.

Although sporadic reports of the presence of chloroform and other trihalomethanes in finished drinking water occurred before 1974, the reports that year by Rook (8) of the Rotterdam Water Utility in the Netherlands and by Bellar, Lichtenberg, and Kroner (9) of the U.S. Environmental Protection Agency (USEPA) clearly demonstrated that these contaminants are formed during the water treatment processs as a result of chlorination. This finding prompted a survey (10) in early 1975 of 80 water utilities in the United States, 79 of which practice free residual chlorination or combined residual chlorination. This survey, the National Organics Reconnaissance Survey (NORS), showed that all of the water utilities that used free chlorine in their treatment practice had varying concentrations of at least one of four trihalomethanes in their finished drinking water and that they were formed during treatment. Follow-up studies in 1975-1977, including the National Organics Monitoring Survey (NORS) (11) supported these conclusions with analyses of samples collected at 113 locations during three different seasons.

These surveys, combined with like results from all over the world, showed that the reaction of chlorine to produce trihalomethanes was widespread and surely had been occurring for as long as chlorine had been in use in water treatment. Since the discovery of trihalomethanes in drinking water, very large efforts have focused on gaining an understanding of natural factors that influence the formation of trihalomethanes, and which lead to means for their control during drinking water treatment.

The four trihalomethanes most commonly found in finished drinking water are chloroform ($CHCl_3$), bromodichloromethane ($CHBrCl_2$), dibromochloromethane

($CHBr_2Cl$), and bromoform ($CHBr_3$). A fifth, dichloroiodomethane ($CHCl_2I$) is found much less frequently. A source of organic precursors that reacts with chlorine is generally accepted to be aquatic humic materials found in virtually all drinking water sources, especially surface supplies. These humic materials include both humic and fulvic acids which, when present in high concentration, account for much of the color in those waters. Natural bromide and iodide are considered to be the non-chlorine halogens which are incorporated into the trihalomethanes.

TRIHALOMETHANES FROM CHLORINATION

The concentrations of trihalomethanes observed in water treatment and distribution are heavily dependent on reaction time, humic material concentration, pH, temperature, and to a smaller extent, chlorine dose/residual (3). Formation is strongly dependent on chlorine dose when the reaction is chlorine limited, but little change can be seen in $CHCl_3$ concentration once the demand requirement has been met and the residual maintained as in the example shown (Figure 5).

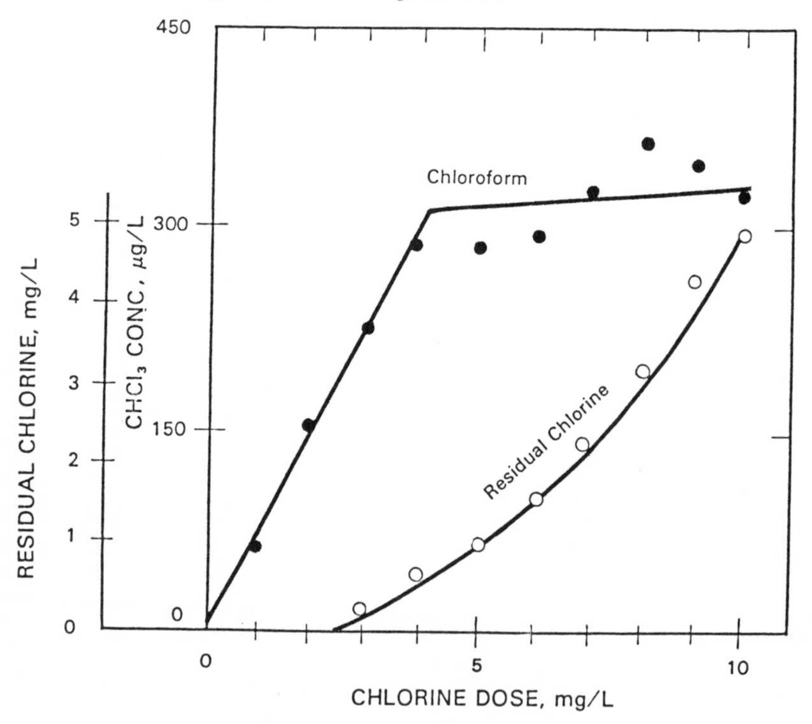

Figure 5. Chloroform formation compared with chlorine residual (from ref. 3, based on ref. 12).

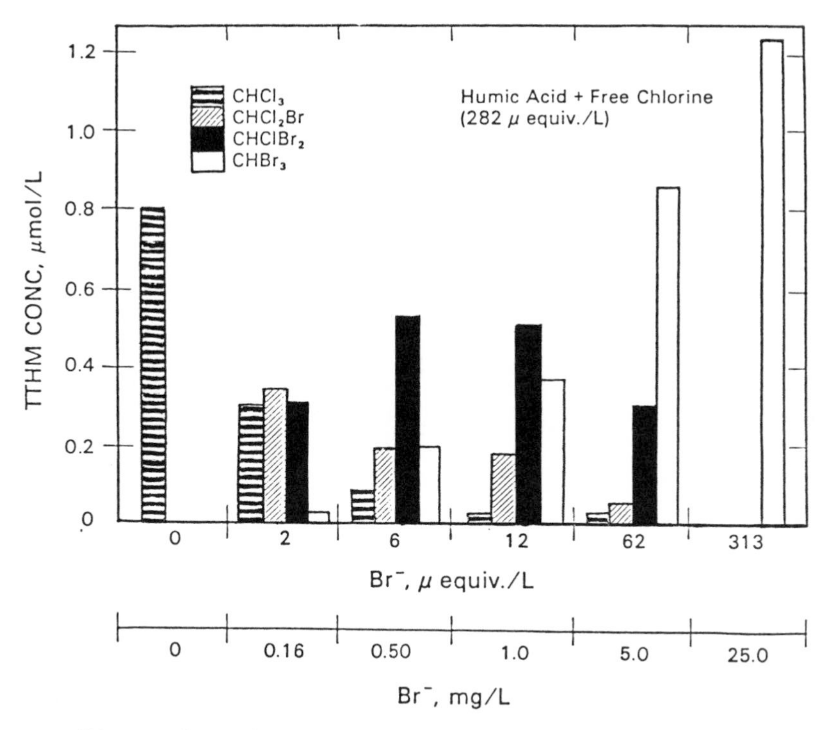

Figure 6. Distribution of trihalomethanes with varying bromide concentration (from ref. 3).

Bromide ion, even at very low concentrations relative to the applied chlorine, strongly influences the mixture of trihalomethanes formed (Figure 6). This occurs as a result of rapid oxidation of Br^- to HOBr followed by reaction of this active species with the organic precursor. The reaction of bromine with humic substances is much more rapid than that of chlorine, resulting in both the disproportionate share of bromine-containing species and approaching a final trihalomethane concentration much earlier (usually minutes rather than hours or days).

After much study of treatment feasibility and careful consideration of potential health effects, the USEPA promulgated a regulation on November 29, 1979 limiting concentrations of total trihalomethanes ($CHCl_3$ + $CHCl_2Br$ + $CHClBr_2$ + $CHBr_3$) to 0.10 mg/L at utilities serving populations greater than 10,000 population (13). This regulation becomes effective over a four year schedule and has caused many utilities to reexamine their treatment processes, including approaches to disinfection. In many cases a change to an alternative disinfection practice either has occurred or is being contemplated by utilities that otherwise have difficulty meeting the maximum contaminant level requirement for trihalomethanes in drinking water.

ALTERNATIVE DISINFECTANTS AND TRIHALOMETHANE FORMATION

Preformed chloramines do not produce trihalomethanes (3). Generally, the addition of ammonia to water containing free chlorine will stop the formation of trihalomethanes. Exceptions to this have been observed, and one possible explanation is that intermediates built-up during the chlorination step are hydrolyzed after the free chlorine is removed with ammonia. The presence of bromide ion also can cause the reaction to apprach completion before the NH_3 is added, reducing the effectiveness of NH_3 as a control measure.

Chlorine dioxide used alone also does not generally form trihalomethanes (14,15), although in one instance trihalomethane formation has been reported at pH 12, very high for drinking water treatment practice (16). Addtionally, chlorine dioxide plus excess chlorine actually results in lower final trihalomethane concentrations than the same amount of chlorine alone (3,14).

Ozone alone does not form trihalomethanes, except possibly in the presence of excessive concentrations of bromide ion (3,4). Ozone can influence the yield of trihalomethanes, either upward or downward, resulting from subsequent chlorination of the ozone-treated water necessary to maintain a disinfectant residual in the distribution system (3).

OTHER REACTIONS

The literature is filled with ample evidence that the active oxidants are sufficiently reactive with organic compounds so that byproduct formation might be expected (1-5). For example, the formation of chlorophenols by chlorine accompanied by the associated odor problems has long been known in the drinking water field. This can be prevented by use of chlorine dioxide, even though chlorophenols can be formed under certain conditions (Table I) and indicates a different kind of reactivity of ClO_2.

As mentioned before, ozone is a very reactive compound and has a very short half-life. Ozone is well known by chemists engaged in organic synthesis to react readily with certain classes of organic compounds.

Chloramines, much less reactive than chlorine, persist for long periods in treated water and probably are less prone to produce identifiable byproducts.

The reactions of all of these oxidants with aquatic humic materials are only beginning to be understood,

TABLE I. PRODUCTS RESULTING FROM CHLORINE DIOXIDE TREATMENT OF PHENOL (15)[a]

	Yield from Phenol%					
ClO_2/ phenol (mole/) (mole[b])	o-Chloro-phenol	Phenol recovered	2,4-Dichloro-phenol	p-Chloro-phenol	Hydro-quinone	Total Recovery
4/5	11	30	0.3	13	3.6	58
14/5	NF[c]	NF	NF	NF	7.2	7.2
14/1	NF	NF	NF	NF	45	45

[a]Reaction time, 4 hr.

[b]In mg/L., 4/5 = 43.5/75; 14/5 = 150/75; 14/1 = 164/16.

[c]None found.

however, primarily because of the complexity and relatively uncharacterized nature of the substrate material itself. Much of this is non-volatile, high molecular weight material that cannot be identified or measured by gas chromatographic techniques (17). Many of the expected byproducts would be expected to fall within this category as well.

TOTAL ORGANIC HALOGEN (TOX)

One approach to characterizing and quantifying byproduct formation from the disinfectants that contain halogen is to measure Total Organic Halogen (TOX) which can be done by use of an activated carbon adsorption-pyrolysis technique (18). In our laboratory, this analytical technology was applied, with appropriate control solutions, to solutions of a soil humic acid that was treated separately with each of the three disinfectants: free chlorine, combined chlorine (chloramines), and chlorine dioxide. Time, temperature, pH, disinfectant dose, and organic substrate concentrations were varied. Selected examples of the results obtained are presented in Figures 7 and 8.

The bar graphs in Figure 7 represent Non-Purgeable Organic Halogen (NPOX) above the zero line and Purgeable Organic Halogen (POX), measured as trihalomethanes (in this

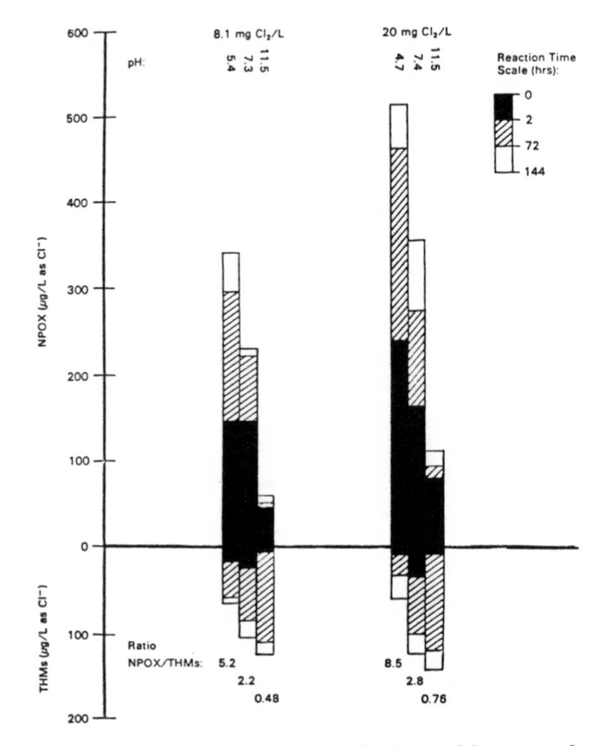

Figure 7. A comparison of the effects of pH and oxidant dose on the formation of NPOX and THMs ($CHCl_3$) at 20° in distilled water solutions of 5 mg humic acid/L. Note that THMs + NPOX = TOX (the entire bar).

case exclusively $CHCl_3$), below the zero line. The full length of a bar, therefore, represents Total Organic Halogen (TOX). Arranged in this manner, the concentration of TOX in solution at 20° C can be seen to be composed mostly of NPOX at pH values of 5 and 7, while at pH 11 the amount of trihalomethanes present is the greater. The ratio of NPOX to trihalomethanes is shown to diminish with increasing pH, which occurs because increasing pH favors a slight increase in trihalomethane formation, while at the same time promoting a drastic decrease in NPOX concentration.

The effect on organic halide formation of increasing the dose of chlorine from 8.1 mg/L to 20 mg/L at 20°C also is demonstrated in Figure 7. Compared with the low dose, TOX is increased by 29, 36 and 27% at pH values near 5, 7 and 11, respectively, at the higher dose. This is almost entirely due to an increase in NPOX formation.

Figure 7 also reveals a relatively rapid formation of NPOX within the first two hours after dosing with chlorine when compared to THM formation. All of these results of chlorination are typical of those reported for humic materials from other sources (19).

Figure 8 compares the NPOX and trihalomethane formation at 20° C at three different pH levels for each of the three disinfectants, chlorine, chloramines, and chlorine dioxide. NPOX formation is reduced by 85% when comparing the use of chloramine as the disinfectant to chlorine, and is even much lower when chlorine dioxide is used. Trihalomethane formation was reduced by greater than 95% when chloramines were used, and no trihalomethanes were detected when chlorine dioxide was used. Unlike the results obtained with chlorine, which increases the formation of trihalomethanes with increasing pH, the use of chloramines and chlorine dioxide as disinfectants results in a decrease in the formation of all organic halides with increasing pH. NPOX formation with chlorine dioxide was rapid with no difference observed between 2 hr and 144 hr sample concentrations.

As part of the above studies, these tests also were run at three temperatures (4°, 20° and 36° C), the general effect of increasing temperatures was to produce more total organic halide whatever the disinfectant used and regardless of the pH. The most significant increase in TOX was always encountered at pH near 5 and was primarily due to an increase in NPOX. In a similar set of tests which involved a lowering of organic substrate concentration from 5 mg/L to 2 mg/L and an increase in chlorine concentration from 8.1 to 10 mg/L, TOX formation was reduced by 55%. A reduction in both THM and NPOX formation occurred in this test.

Thus, the evidence from varying chlorine disinfectant concentration and organic substrate concentration indicates that at any given pH and temperature, THM formation increases principally as a function of organic substrate concentration, and NPOX formation increases as a function of both organic substrate concentration and chlorine concentration.

OTHER SPECIFIC COMPOUNDS PRODUCED

Specific compounds which account for a share of the observed byproducts comprising the total organic halogen other than trihalomethanes are being found. These include dihaloacetonitriles and mono-, di-, and trichloroacetic acids. Numerous other compounds, some containing chlorine, have been identified with relatively confident structural assignment (20,21). For the most part, however, this work has involved experiments designed to maximize byproduct yield, rather than mimic the drinking water treatment practice. Part of this work is the subject of a paper to be presented later this afternoon by Norwood (22).

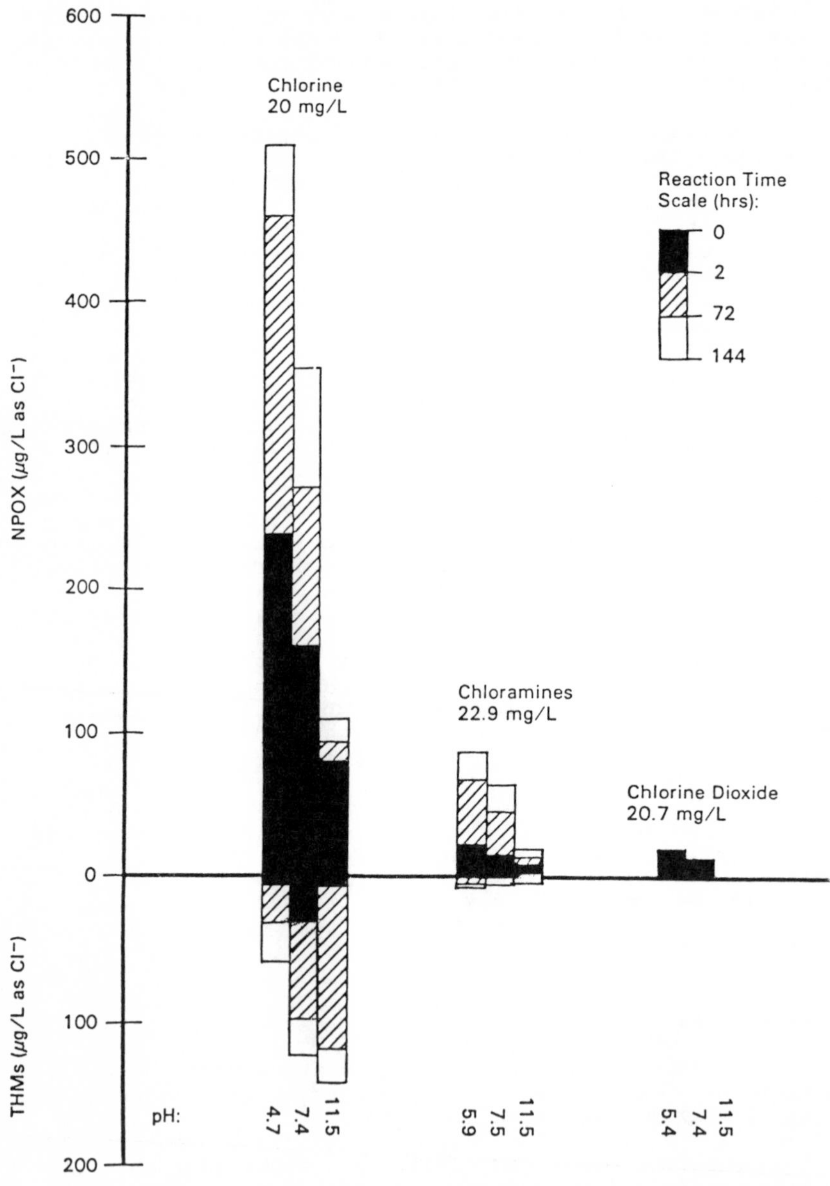

Figure 8. A comparison of the formation of NPOX and THMs ($CHCl_3$) at 20° in distilled water solutions of 5 mg humic acid/L dosed with various disinfectants. Note that NPOX + THMs = TOX (the entire bar).

Figure 9 shows computerized reconstructions of fused silica capillary column gas chromatograph/mass spectrometry data from 13 L of unchlorinated and 10 L of chlorinated (20 mg/L Cl_2) solutions of a soil humic acid (5.3 mg/L TOC). In unpublished studies conducted in our laboratory, the samples were concentrated on granular activated carbon, extracted with ethyl ether, and methylated with diazomethane. Forty nanograms of 1-chlorododecane was injected with each sample, and each chromatogram has been adjusted for a 100 percent response for the internal standard. This example graphically illustrates in a qualitative way (17) the chromatographable byproducts produced by the addition of chlorine to this sample.

Similar earlier studies in our laboratory, using XAD-8 resin to concentrate the chlorination byproducts attempted to compare the same soil humic acid with an aquatic humic acid and a finished water from a surface water with a relatively low concentration of humic material. The soil humic acid experiment was performed twice to check the reproducibility of the methodology, and approximately twice the volume of tap water was concentrated because of the low humic material concentration. Both mass spectral data and chromatographic retention time data were used to check for recurrence of individual compounds. Actual identification was not attempted in most cases, and most spectra did not match available reference spectra. Designation of a peak as chlorinated was based on visual inspection of mass spectral data for characteristic ion clusters.

Table II lists the numbers of chlorination byproducts (unique retention time and spectra), both chlorinated and nonchlorinated, from various humic materials. Trihalomethanes and compounds also found in the nonchlorinated samples or solvent blanks are not included in the list. Of all the compounds detected in the three samples, the soil humic acid produced the most byproducts, especially chlorinated compounds. The aquatic humic acid produced approximately half as many as the soil, and the finished tap water data indicated that, if produced from this source, relatively few of the byproducts are surviving the treatment process. The soil humic acid had the largest number of byproducts that were not found in the other samples. In contrast, only three compounds were unique to the tap water sample, indicating that those that did form were typical of aquatic humic material byproducts.

INORGANIC END PRODUCTS

Each of the active oxidants may produce reduced species or disproportionation products of concern. Free chlorine is mostly dissipated as chloride ion at concentrations which are insignficant relative to normal background levels.

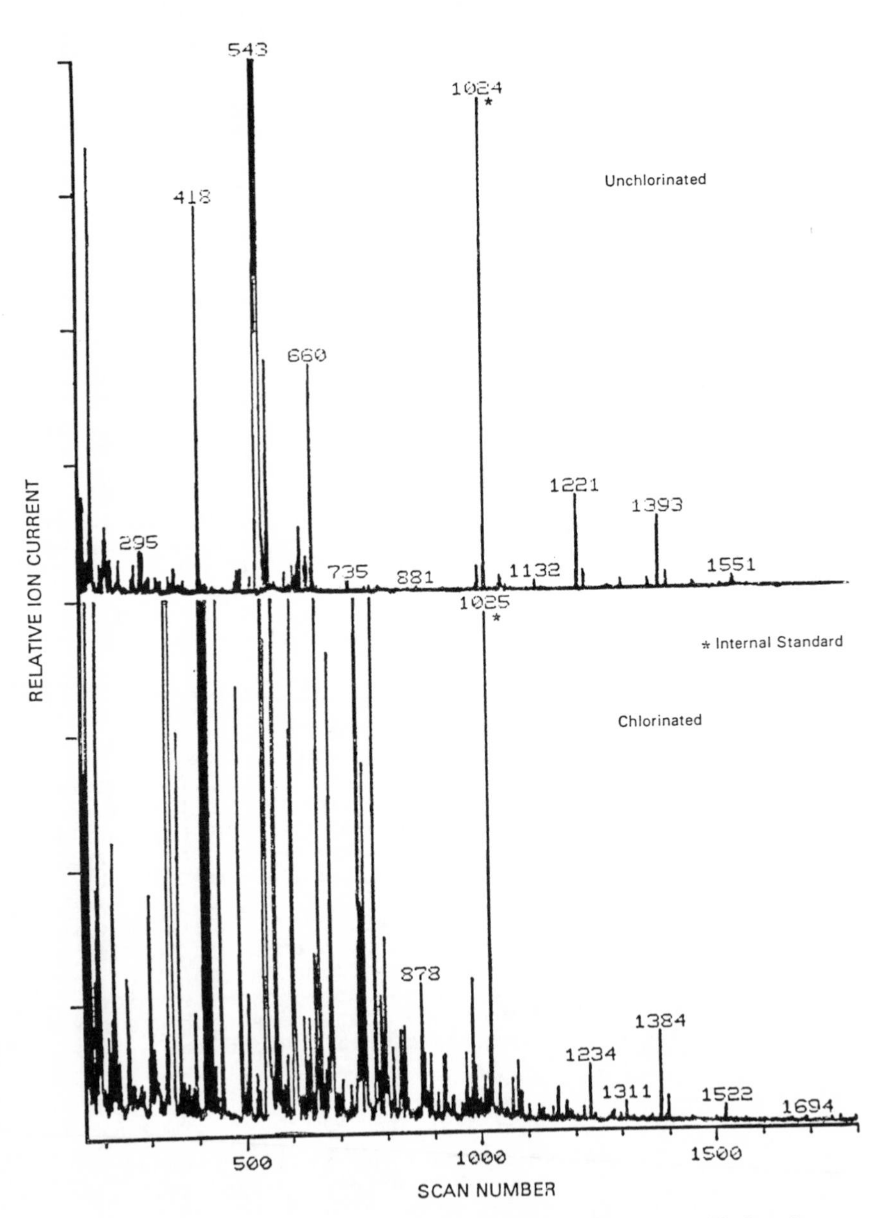

Figure 9. Chlorinated and unchlorinated soil humic acid total ion current profiles.

TABLE II. BYPRODUCTS FOUND FROM CHLORINATION OF DIFFERENT TYPES OF HUMIC ACID SOURCES

	Number of Organics Found	
	Chlorine-containing	Nonchlorine containing
Total Unique Compounds (All Samples)	47	51
Soil Humic Acid	43	37
Aquatic Humic Acid	27	24
Finished Water	8	17
Unique to Soil Humic Acid	16	18
Unique to Aquatic Humic Acid	4	11
Unique to Finished Water	0	3

When chlorine dioxide reacts with organic compounds the byproduct chlorite (ClO_2^-) is formed. Futhermore, as chlorine dioxide disproportionates in water, both chlorite and chlorate (ClO_3^-) are formed as byproducts. The relative proportion of these byproducts was determined during a study reported elsewhere by Miltner (23), in which 1.5 mg/L of chlorine dioxide was added to Ohio River water that had been treated in a pilot plant. The data in Table III show that approximately 50% of the original chlorine dioxide was converted to chlorite, about 25% to chlorate, and approximately 25% to chloride. Thus, when chlorine dioxide is used as an alternative disinfectant, the health significance of inorganic anions other than chloride must be considered. These inorganic byproducts are unique to chlorine dioxide.

Chloramines themselves persist in a drinking water distribution system, and have been known to cause unique problems in kidney dialysis and to tropical fish (3).

SUMMARY

Considerations such as disinfection power, ease of application, and low cost in the past have led to the use of free chlorine ($HOCl/OCl^-$) as the primary disinfectant in drinking water treatment. Discovery of trihalomethanes, formed by the action of free chlorine upon natural organic materials such as aquatic humic materials, has led to a reexamination of this practice. In many cases a change to an alternative disinfection practice, such as the use of

TABLE III. INORGANIC CHLORINE DIOXIDE BYPRODUCTS[a] (14,23)

	Initial Concentration		Final Concentration		
Species	mg/L	mg/L as Cl^-	mg/L	mg/L as Cl^-	ClO_2 demand, %
ClO_2	1.5	0.8	0	0	-
ClO_2^-	-	-	0.7	0.4	50
ClO_3^-	-	-	0.4	0.2	25
Cl^-	17.9	17.9	18.1	18.1	25
Totals	-	18.7	-	18.7	100

[a]1.5 mg/L ClO_2 added to Ohio River water which was coagulated, settled, dual-medial filtered. Contact time was 42 hr; pH was 7.1.

chloramines, chlorine dioxide, or ozone either has occurred or is being contemplated by utilities that otherwise have difficulty meeting the maximum contaminant level requirement for trihalomethanes. These alternative disinfectants when used alone do not usually cause significant formation of trihalomethanes. Each is an active oxidant, however, and has the potential of forming objectionable byproducts other than trihalomethanes (as may chlorine as well). Free chlorine, for example, forms "organic halogen" levels substantially in excess of those accounted for by the trihalomethanes. Chloramines and chlorine dioxide do also, although less than does free chlorine. Increasing numbers of specific reaction products of each disinfectant in aqueous solution now are being observed, and some are being identified. Thus, the benefical use of each of the chemicals as a biocide in drinking water treatment may have its own less desirable side effects.

REFERENCES

1. Jolley, R.L. and Carpenter, J.H., "A review of the Chemistry and Environmental Fate of Reactive Oxidant Species in Chlorinated Water", In: Water Chlorination, Environmental Impact and Health Effects, Volume 4, Book 1. R. L. Jolley, et al., Eds., (Stoneham, MA; Ann Arbor Science Publishers, Inc., 1983), pp. 3-48.

2. Gordon, G., Kieffer, R.G., and Rosénblatt, D.H., "The Chemistry of Chlorine Dioxide", In: Progress in Inorganic Chemistry, Vol. 15, S.J. Lippard, Ed. New York, NY; Wiley-Interscience, 1972) p. 201.
3. Symons, J.M., Stevens, A. A., Clark, R.M., Geldreich, E.E., Love, O.T. Jr., and DeMarco, J., "Treatment Techniques for Controlling Trihalomethanes in Drinking Water." (EPA Report No. 600/2-81-156) (Cincinnati, OH; US EPA, 1981).
4. Rice, R.G., and Cotruvo, J.A., Eds. Ozone/Chlorine Dioxide Oxidation Products of Organic Materials, (Norwalk, CT; Intl Ozone Assoc., 1978).
5. Miller G.W., Rice, R.G., Robson, C.M., Scullin, R.L., Kühn, W. and Wolf, H., "An Assessment of Ozone and Chlorine Dioxide Technologies for Treatment of Municipal Water Supplies," EPA Report No. 600/2-78-147 (Cincinnati, OH; USEPA August 1978), 571 pp., NTIS Accession No. PB 285972/AS.
6. Scarpino, P.V., Cronier, S., Zink, M.L., Brigano, F.A.O., and Hoff, J.C., "Effects of Particulates on Disinfection of Enteroviruses and Coliform Bacteria in Water by Chlorine Dioxide," In: Proceedings - Fifth Water Quality Technology Conference, Kansas City,
"The Chemistry of Chlorine Dioxide", In: Progress in Inorganic Chemistry, Vol. 15, S.J. Lippard, Ed. New York, NY; Wiley-Interscience, 1972) p. 201.
3. Symons, J.M., Stevens, A. A., Clark, R.M., Geldreich, E.E., Love, O.T. Jr., and DeMarco, J., "Treatment Techniques for Controlling Trihalomethanes in Drinking Water." (EPA Report No. 600/2-81-156) (Cincinnati, OH; US EPA, 1981).
4. Rice, R.G., and Cotruvo, J.A., Eds. Ozone/Chlorine Dioxide Oxidation Products of Organic Materials, (Norwalk, CT; Intl Ozone Assoc., 1978).
5. Miller G.W., Rice, R.G., Robson, C.M., Scullin, R.L., Kühn, W. and Wolf, H., "An Assessment of Ozone and Chlorine Dioxide Technologies for Treatment of Municipal Water Supplies," EPA Report No. 600/2-78-147 (Cincinnati, OH; USEPA August 1978), 571 pp., NTIS Accession No. PB 285972/AS.
6. Scarpino, P.V., Cronier, S., Zink, M.L., Brigano, F.A.O., and Hoff, J.C., "Effects of Particulates on Disinfection of Enteroviruses and Coliform Bacteria in Water by Chlorine Dioxide," In: Proceedings - Fifth Water Quality Technology Conference, Kansas City, MO, December 4-7, 1977, Paper 2B-3, 11 pp. (Denver, CO, Am. Water Works Assoc., 1978).

7. Eposito, M.P., "The Inactivation of Viruses in Water by Dichloramine," M.S. Thesis, University of Cincinnati, Cincinnati, OH (1944).
8. Rook, J.J., "Formation of Haloforms During Chlorination of Natural Water," Water Treatment and Examination, 23(2):234-243 (1974).
9. Beller, T.A., Lichtenberg, J.J., and Kroner, R.C., "The Occurence of Organohalides in Chlorinated Drinking Water," J. Am. Water Works Assoc. 66(12):703-706 (1974).
10. Symons, J.M., Bellar, T.A., Carswell, J.K., DeMarco, J., Kropp K.L., Robeck, G.G., Seeger, D.R., Slocum, C.J., Smith B.L., and Stevens, A.A., "National Organics Reconnaissance Survey for Halogenated Organics," J. Am. Water Works Assoc. 67(11):634-647 (1975).
11. Brass, H.J., Feige, M.A., Halloran, T., Mello, J.W., Munch, D., and Thomas, R.F., "The National Organic Monitoring Survey: Sampling and Analysis for Purgeable Organic Compounds," In: Drinking Water Quality Enhancement Through Source Protection, (Stoneham, MA; Ann Arbor Science Publishers, Inc., 1977), p. 393.

CHAPTER 9

BY-PRODUCTS OF CHLORINATION: SPECIFIC COMPOUNDS AND THEIR RELATIONSHIP TO TOTAL ORGANIC HALOGEN

D.L. Norwood, G.P. Thompson, J.J. St. Aubin, D.S. Millington, R.F. Christman, and J.D. Johnson

Department of Environmental Sciences and Engineering
School of Public Health
University of North Carolina
Chapel Hill, North Carolina 27514

ABSTRACT

Over the past decade a great deal of scientific interest has focused on the hydrophobic halogenated by-products of drinking water disinfection, principally chloroform and other trihalomethanes. Natural aquatic humic material has been implicated as a precursor for these substances. More recent data suggests that the importance of chloroform as an aqueous chlorination product of humic materials may have been overstated. This study reviews the progress to date on the identification and quantification of more hydrophilic chlorination by-products, such as trichloroacetic acid, from a variety of humic substances. Qualitative data are summarized from several previous studies and compared with more recent experiments performed in the authors' laboratory on isolated aquatic humic material. Data are reported which support the hypothesis that these hydrophilic substances are produced in greater yield than chloroform and account for a significant fraction of the total organic halogen materials produced.

INTRODUCTION

Over the past decade a great deal of scientific interest has focused on the hydrophobic halogenated by-products of drinking water disinfection, principally chloroform and its sister trihalomethanes. Aquatic humic materials which are present in most natural waters have been implicated as major precursors for these substances. However, it does not seem reasonable to assume that this relatively hydrophilic humic material should produce mostly hydrophobic material (i.e., chloroform) when oxidized by aqueous chlorine. On the contrary, one would anticipate the production of even more hydrophilic substances. This hypothesis has been the subject of a number of studies in our laboratory and by others utilizing different natural waters and humic substances.

In an early series of experiments, Oliver (1) studied the total amount of organic chlorine that was incorporated when various isolated humic materials were reacted with chlorine under typical water treatment plant conditions. This study utilized a new group parameter termed Total Organic Halogen (TOX). In all cases, it was found that considerable nonvolatile organic halogen material was produced, which on the average accounted for 43% of the total. Volatile organic halogen material also was produced, of which 95% was found to be chloroform by independent measurements. Also, the relative amounts of volatile versus nonvolatile organic halogen were pH-dependent, with the nonvolatile fraction being maximized at more acidic values. Although Oliver identified none of the specific compounds which comprised the nonvolatile organic halogen fraction, it is attractive to assume that the relative lack of volatility was caused by increased polarity and thus hydrophilicity.

Glaze and co-workers have reported similar results for chlorinated raw water samples (2,3). In one study, chlorinated raw water of high organic carbon content was subjected to size fractionation by high performance liquid chromatography in conjunction with TOX analysis (2). The majority of the TOX appeared as apparently higher molecular weight species which maximized at about M = 4.2 K.

Later results showed that the TOX formation potential of another high carbon raw water sample greatly exceeded the trihalomethane formation potential (3), again pointing to the presence of nonvolatile TOX materials. Size exclusion chromatograms of this chlorinated raw water before and after passage through a column of XAD-2 resin indicated that this material was very inefficient at adsorbing the nonvolatile chlorinated organic material. This result was confirmed by Watts, et al (4). Since the XAD-2 adsorption effect is mainly due to weak intermolecular bonding forces, it is again attractive to assume that this decrease in adsorbability was due to increased polarity and hydrophilicity.

These studies represent only a small sampling of the large body of literature indicating that trihalomethane formation alone cannot account for the majority of TOX production from aqueous natural product organic materials. Futhermore, it has been demonstrated that the mutagenic activity of chlorinated drinking water is more closely associated with this nonvolatile fraction (5,6).

Two hypotheses may be put forward regarding the identity of compounds responsible for this nonvolatile TOX. First, they may be high molecular weight macromolecular species, and thus not be amendable to the usual methods of chemical analysis. Glaze, et al (2,3) and Watts et al. (4) allude to this possibility. Secondly, it is possible that the compounds are not of extremely high molecular weight, but are quite polar and hydrophilic, and thus would require special procedures for their extraction and analysis.

The second hypothesis, which was stated at the beginning of this section, has been the subject of several recent reports in the literature. A variety of small aliphatic chlorinated hydrophilic compounds was determined in our laboratory to be produced from chlorination of an extracted aquatic humic material under various reaction conditions (7-9). A summary of products common to all experiments is shown in Table I. The dominant structures found in each case were dichloroacetic acid (DCAA) and trichloroacetic acid (TCAA).

Other workers have shown that a similar series of hydrophilic products was produced from chlorination of extracted soil humic material with the same dominance of DCAA and TCAA (10-12). In these studies the product distribution was shown to be a function of several variables, including pH and the mole ratio of chlorine to carbon (12).

Concurrently it was shown by Rook that DCAA and TCAA were the principal constituents of a methylene chloride extract of Rotterdam finished drinking water subjected to breakpoint chlorination (13). Also, Boyce and Horning (14) showed that a significant amount of chloroform was produced and detected when aqueous TCAA was injected into a gas chromatograph. This chloroform, presumably produced by hydrolysis in the GC injection port, may be responsible for excess chloroform reported from direct aqueous injection analytical methods and attributed to "chloroform precursors" (15).

In all the above mentioned studies on aquatic and soil humic substances it was shown that TCAA was produced in greater quantities than chloroform. These results beg the question: how much TOX is represented by these hydrophilic compounds? In our laboratory, we initially studied product and TOX production from an extracted aquatic fulvic acid in highly concentrated solution (16). The principal products were found in abundances indicated in Table II. The sum of these products accounts for a significant fraction (53%) of the TOX. Research now is in progress in our laboratory on natural water samples to determine the product distributions and relative amounts. Initial results of these studies are presented here.

TABLE I. SHORT CHAIN CHLORINATION PRODUCTS OF AQUATIC HUMIC AND FULVIC ACIDS

Formula	Name
$CHCl_3$	trichloromethane (chloroform)
$CHBrCl_2$	bromodichloromethane
CCl_3CHO	trichloroacetaldehyde (chloral)
H_2CClCO_2H	chloroethanoic acid (chloroacetic acid)
$HCCl_2CO_2H$	dichloroethanoic acid (dichloroacetic acid, DCAA)
CCl_3CO_2H	trichloroethanoic acid (trichloroacetic acid, TCAA)
$CH_3CCl_2CO_2H$	2,2-dichloropropanoic acid
$CCl_2{=}CHCO_2H$	3,3-dichloropropenoic acid
$CCl_2{=}CClCO_2H$	2,3,3-trichloropropenoic acid
$HO_2CCCl_2CO_2H$	dichloropropanedioic acid (dichloromalonic acid, DCMA)
$HO_2C(CH_2)_2CO_2H$	butanedioic acid (succinic acid)
$HO_2CCH_2CHClCO_2H$	chlorobutanedioic acid (chlorosuccinic acid)
$HO_2CCCl_2CH_2CO_2H$	2,2-dichlorobutanedioic acid (2,2-dichlorosuccinic acid, (DCSA)
$HO_2CCH{=}CClCO_2H$	*cis*-chlorobutenedioic acid (chloromaleic acid)
$HO_2CCCl{=}CClCO_2H$	*cis*-dichlorobutenedioic acid (dichloromaleic acid)
$HO_2CCCl{=}CClCO_2H$	*trans*-dichlorobutenedioic acid (dichlorofumaric acid)

TABLE II. YIELDS OF CHLORINATION PRODUCTS FROM EXTRACTED FULVIC ACID (BLACK LAKE)

Product	mg/g FA	mg C/g FA	% orginal TOC	% final TOX
trichloroacetic acid	90.3	13.3	3.0	32.1
$CHCl_3$	38.2	3.8	0.8	17.3
dichloroacetic acid	10.2	1.9	0.4	3.6*
dichlorosuccinic acid	3.4	0.9	0.2	--
TOTAL	142.1	19.9	4.4	53.0

* Sum of dichloroacetic and dichlorosuccinic acids

EXPERIMENTAL

Natural water samples for organic halide screening were collected in acid washed 1-L glass bottles and stored until chlorination at room temperature in a head-space free condition. Samples of each water source also were collected in acid washed 60-mL septum-capped vials for total organic carbon (TOC) analysis utilizing a Dohrmann DC-80 instrument equipped with an automatic sampler.

Chlorination reactions were carried out by adding a sufficient quantity of 5% sodium hypochlorite solution to each 1-L sample to produce a chlorine to carbon mole ratio of 4:1. Head-space free conditions and neutral pH were maintained for the entire reaction period of 24 hrs. An organic halide blank was prepared by adding 2.0 mL of 5% NaOCl solution to 1-L of distilled-deionized water. After the reaction time period, each sample was checked iodometrically and found to contain residual chlorine. This excess chlorine was quenched by addition of sodium arsenite crystals which had been previously extracted with ether to remove potential contaminants. Aliquots of each quenched reaction mixture were placed in acid-washed septum-capped vials and submitted for total organic halogen (TOX) analysis using EPA Interim Method 450.1 (17) with a Dohrmann MCTS-20 instrument.

Measured 300 mL aliquots of each reaction mixture and blank then were spiked with approximately 50 microg/L of the compounds ^{13}C -trichloroacetic acid and ^{13}C-chloroform. These structures are stable isotopically labeled analogs of the two compounds of interest. After allowing a two hour time period for these internal standards to equilibrate with the sample matrix, each reaction mixture was acidified to pH 0.7 with concentrated HCl and extracted with three 100-mL aliquots of redistilled diethyl ether. The ether extracts were dried with liquid nitrogen and pre-extracted $MgSO_4$, concentrated in a Kuderna-Danish apparatus and methylated with ethereal diazomethane. This sample work-up procedure is identical to that described previously (16). The generalized sequence of steps is outlined in Figure 1.

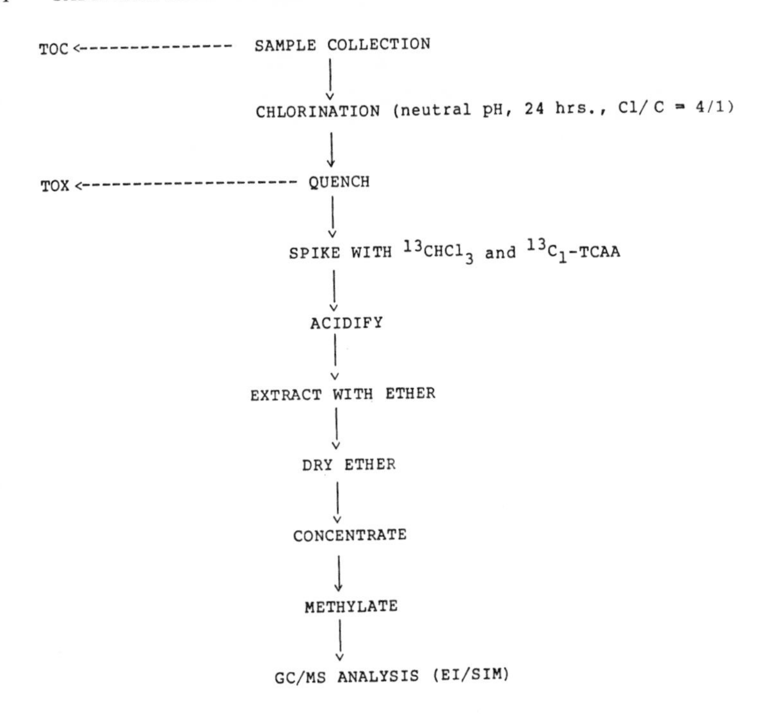

Figure 1. Outline of steps in the sample work-up and analysis procedure.

Each concentrated, methylated ether extract was analyzed by capillary column gas chromatography/mass spectrometry (GC/MS) utilizing an isotope dilution method to quantify the amounts of trichloroacetic acid and chloroform produced during chlorination. The isotopically labeled internal standards spiked into the aqueous samples should correct for matrix effects and recovery losses during work-up. The method utilizes procedures similar to those outlined in EPA Method 1625 (18), and has been described previously (16). The GC/MS/DS consists of a VG-Micromass 7070F double-focusing mass spectrometer with VG 2035 computer interfaced to an HP5710A capillary gas chromatograph. Data acquisition was under computer control with selected ion monitoring (SIM) of the quantitation ion pairs m/z 85 and 86 for chloroform and m/z 141 and 142 for methyl trichloroacetate. Detailed descriptions of the isotope dilution methodology, instrumental conditions and validation have been described in detail elsewhere (19).

RESULTS AND DISCUSSION

The selection of water sources for the organic halide screening experiment included examples from three major types of North Carolina waters. The first group represented is the so-called "Carolina bay lakes" which are highly colored, high TOC, natural lakes located on the coastal plain. These included Lake Singletary, Lake Waccamaw and Black Lake. Isolated Black Lake fulvic acid had been shown previously to yield large quantities of hydrophilic chlorinated reaction products and chloroform (see Table II).

The second group represented impoundments which are used as drinking water supplies, and included Lake Wheeler (Raleigh, NC), Lake Michie (Durham, NC) and University Lake (Chapel Hill, NC). Finally, the third group represented rivers which also are used as drinking water supplies, and included the Haw River (Pittsboro, NC) and the Cape Fear River sampled at two locations (Sanford and Fayetteville, NC). These sources represent a reasonable cross-section of North Carolina natural waters.

Before screening all the samples for organic halide, it was necessary to determine what sort of reaction conditions to employ in order to produce reasonable quantities of these substances to measure. To accomplish this, Black Lake was selected as the baseline source (since so much data on it already was in hand) and samples were chlorinated at various chlorine to carbon molar ratios between 0.5 and 4. The results of the experiment are shown in Table III.

It should be noted first that significant quantities of TCAA, chloroform and TOX were produced in these reactions, with the sum of the TCAA and chloroform accounting for a significant fraction of the TOX (59% at Cl/C = 4). This

TABLE III. RESULTS OF CHLORINATION OF BLACK LAKE WATER VARYING Cl/C

Sample	Cl/C Mole Ratio	TOX (microg/L as Cl)	TCAA microg/L	$CHCl_3$ microg/L	% TOX Accounted for
1	0.5	560	77.2 (50.2)[+]	39 (34.7)	15
2	1.0	970	244 (159)	113 (101)	27
3	2.0	940	278 (181)	258 (230)	43
4	4.0	1000	299 (194)	295 (263)	59

[+] microg/L as Cl indicated by values in parenthesis

result is in good agreement with previous data (16). It should be noted also that all three of these concentrations appear to be leveling off at a chlorine to carbon mole ratio of 4/1; therefore this value was chosen as the condition for chlorination of all the other samples.

The results of the complete screening analysis are presented in Table IV. Again, it should be noted that significant quantities of TCAA, chloroform and TOX were produced from all the sources studied. In this context, significant is defined in terms of the EPA 100 microg/L standard for chloroform (20). Of special interest is the observation that in all cases TCAA was at least as concentrated as chloroform, and in many instances more so. The sum of these two components makes up a significant fraction of the TOX in all cases. Predictably, the highly colored bay lakes Singletary and Waccamaw produced the most TCAA, chloroform and TOX. Large quantities also were produced in the impounded lakes and rivers used as drinking water supplies.

It is appropriate at this point to comment on the recovery of TOX by the EPA Interim Method 450.1 utilized in these studies. This is important since the data on percent of TOX accounted for assumes quantitative recovery of all organic halides present, especially TCAA and chloroform. Studies have been reported implying that the method is quite

TABLE IV. RESULTS OF CHLORINATION OF VARIOUS NORTH CAROLINA NATURAL WATERS

Sample	Source	TOC microg/L	TOX microg/L	TCAA microg/L	$CHCl_3$ microg/L	%TOX Accounted for
1	Lake Waccamaw	8640	3280	1370 (892)*	653 (582)	44.9
2	Lake Singletary	6580	3440	1620 (1050)	834 (743)	52.1
3	Lake Wheeler	5170	1700	545 (355)	291 (259)	36.1
4	Cape Fear River, Sanford	4770	990	410 (267)	281 (250)	52.2
5	Lake Michie	4280	1430	617 (402)	331 (295)	48.7
6	Black Lake	3770	940	348 (227)	346 (308)	56.9
7	Cape Fear River, Fayetteville	3610	1000	392 (255)	285 (254)	50.9
8	Haw River Pittsboro	3100	825	315 (205)	216 (192)	48.1
9	University Lake	1440	460	124 (80.7)	93.8 (83.6)	35.7

* microg Cl/L indicated by values in parenthesis

efficient at TOX recovery (17), however, matrix effects on recovery also have been observed (21). Standard addition experiments in our laboratory have shown that TCAA is approximately 93% recoverable as TOX and chloroform approximately 75% recoverable. These standard additions were made to chlorinated Black Lake fulvic acid matrix. Although some matrix effects appear to be present, we feel that the % TOX accounted for data does represent a reasonable basis for comparison at this point. Experimentation is continuing in this area.

Some interesting relationships across the various sources have been observed and are depicted in Figures 2 and

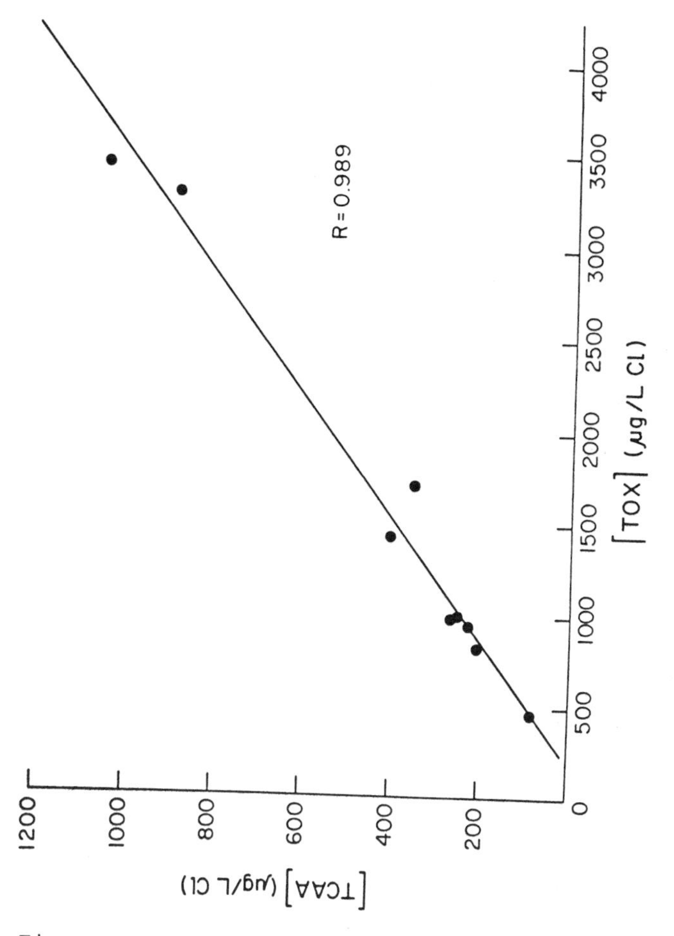

Figure 2. Relationship between trichloroacetic acid and total organic halide.

3. In Figure 2 note that there is a very good linear correlation between TCAA and TOX produced, which implies that this compound represents a relatively constant fraction of the total organic halogen across the various sources. This relationship is not quite so good for chloroform and TOX,, as may been seen in Figure 3, but there is still a good correlation. These relationships seem to imply that TCAA concentration is a better indicator of the total organic halogen present in a chlorinated raw water sample than chloroform. However, great care must be exercised in this interpretation, since only one set of reaction conditions and only a few sources were employed.

CONCLUSIONS

The results of this study indicate that the hydrophilic organic halide trichloroacetic acid is a major by-product of natural water chlorination under the conditions utilized. In many cases it was produced in higher concentrations on a weight basis than chloroform. When summed together, these two structures also can account for a large fraction of the measured total organic halogen. These results clearly indicate the need for monitoring of the hydrophilic portion of the TOX in drinking water.

ACKNOWLEDGEMENT

This research was supported in part by EPA Research Grant No. R804430 from the Municipal Environmental Research Laboratory, Cincinnati, OH, Alan A. Stevens, Project Officer.

REFERENCES

1. Oliver, B.G. "Chlorinated Non-Volatile Organics Produced by the Reaction of Chlorine and Humic Materials", Canadian Res. 11(6):21-22 (1978).
2. Glaze, W.H. and G.R. Peyton, "Soluble Organic Constituents of Natural Waters and Wastewaters Before and After Chlorination", in Water Chlorination: Environmental Impact and Health Effects, Vol. 2, R.L. Jolley, et al., Eds. (Stoneham, MA; Ann Arbor Science Publishers, Inc., 1978), pp. 3-14.

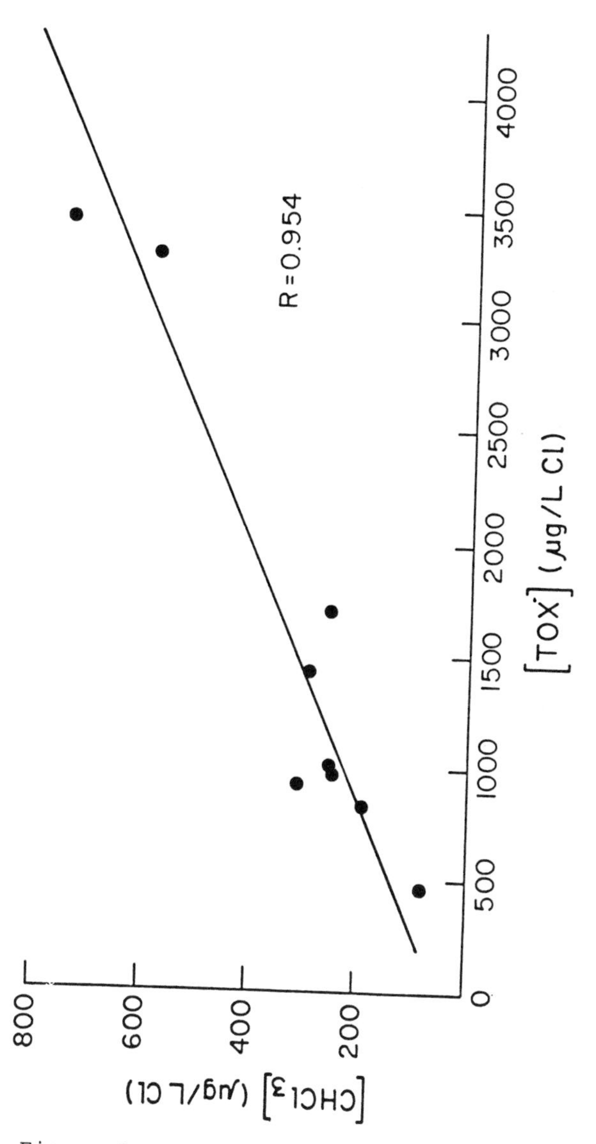

Figure 3. Relationship between chloroform and total organic halide.

3. Glaze, W.H., F.Y. Saleh and W. Kintsley, "Characterization of Nonvolatile Halogenated Compounds Formed During Water Chlorination", in Water Chlorination: Environmental Impact and Health Effects, Vol. 3, R.L. Jolley, et al., Eds. (Stoneham, MA; Ann Arbor Science Publishers, Inc., 1980), pp. 99-108
4. Watts, C.D., B. Crathorne, M. Fielding and S.D. Killops, "Nonvolatile Organic Compounds in Treated Waters", Environ. Health Perspectives 46:87-99 (1982).
5. Bull, R.J. "Health Effects of Drinking Water Disinfectants By-Products", Environ. Sci. Technol. 16(10):554A (1982).
6. Kool, H.J., C.F. van Kreijl, E. de Greef and H.J. van Kronen, "Presence, Introduction and Removal of Mutagenic Activity During Preparation of Drinking Water in the Netherlands", Environ. Health Perspectives 46:207-211 (1982).
7. Christman, R.F., J.D. Johnson, F.K. Pfaender, D.L. Norwood, M.R. Webb, J.R. Hass, and M.J. Bobenreith. "Chemical Identification of Aquatic Humic Chlorination Products", in Water Chlorination: Environmental Impact and Health Effects, Vo. 3, R. L. Jolley, et al., Eds. (Stoneham, MA; Ann Arbor Science Publishers Inc., 1980), pp. 75-84.
8. Johnson, J.D., R.F. Christman, D.L. Norwood, and D.S. Millington, "Reaction Products of Aquatic Humic Substances with Chlorine", Environ. Health Perspectives 46:63-71 (1982).
9. Norwood, D.K., J.D. Johnson, D.S. Millington, and R.F. Christman, "Chlorinated Products from Aquatic Humic Material at Neutral pH", in Water Chlorination: Environmental Impact and Health Effects, Vol. 4, R.L. Jolley et al., Eds. (Stoneham, MA; Ann Arbor Science Publishers, Inc., 1983), pp. 191-200.
10. Quimby, B.D., M.F. Delaney, P.C. Uden and R.M. Barnes, "Determination of the Aqueous Chlorination Products of Humic Substances by Gas Chromatography with Microwave Emission Detection", Anal. Chem. 52:259-263 (1980).
11. Coleman, W.E., J.W. Munch, W.H. Kaylor, H.P. Ringhand, and J.R. Meier, "GC/MS Analysis of Mutagenic Extracts of Aqueous Chlorinated Humic Acids -- A Comparison of the By-Products to Drinking Water Contaminants", Am. Chem. Soc., Div. Environ. Chem. 23(1):53-55 (1983).
12. Miller, J.W. and P.C. Uden, "Characterization of Nonvolatile Chlorination Products of Humic Substances", Environ. Sci. and Technol. 17(3): -150-157 (1983).

13. Rook, J.J., "Possible Pathways for the Formation of Chlorinated Degradation Products During Chlorination of Humic Acids and Resorcinol", in Water Chlorination: Environmental Impact and Health Effects, Vol. 3, R.L. Jolley, et al., Eds. (Stoneham, MA; Ann Arbor Science Publishers, Inc., 1980), pp. 85-98.
14. Boyce, S.D. and J.F. Hornig, "Formation of Chloroform from the Chlorination of Diketones and Polyhydroxybenzenes in Dilute Aqueous Solution", in Water Chlorination: Environmental Impact and Health Effects, Vol. 3, R.L, Jolley, et al., eds. (Stoneham, MA; Ann Arbor Science Publishers, Inc., 1980), pp. 131-140.
15. Pfaender, F.K., R.B. Jonas, A.A. Stevens, L. Moor, and J.R. Hass, "Evaluation of Direct Aqueous Injection Method for Analysis of Chloroform in Drinking Water", Environ. Sci. and Technol. 12(4):438-441 (1978).
16. Christman, R.F., D.L. Norwood, D.S. Millington, J.D. Johnson, and A.A. Stevens, "Identity and Yields of Major Halogenated Products of Aquatic Fulvic Acid Chlorination", Environ. Sci. Technol. 17(10):625-628 (1983).
17. "Total Organic Halide: Method 450.1 Interim," U.S. Environmental Protection Agency, Office of Research and Development, Environmental Monitoring and Support Laboratory, Physical and Chemical Methods Branch, Cincinnati, OH (November 1980).
18. "Semivolatile Organic Compounds by Isotope Dilution GC/MS", U. S. Environmental Protection Agency, WH552, Washington, DC.
19. Norwood, D.L., R.F. Christman, D.S. Millington, and J.R. Hass, "A Quantitative Isotope Dilution GC/MS Method for Trichloroacetic Acid: A Major Product of the Aqueous Fulvic Acid HOCl Reaction", presented at First Intl. Meeting, Intl. Humic Substances Society (1983).
20. "National Interim Primary Drinking Water Regulations", Federal Register 44:68624-68707 (1979).
21. Dressman, R.C. and A.A. Stevens, "The Analysis of Organohalides in Water - an Evaluation Update", J. Am. Water Works Assoc. 75:431-434 (1983).

CHAPTER 10

OZONE FOR DRINKING WATER TREATMENT -- EVOLUTION AND PRESENT STATUS

Rip G. Rice, Ph.D.

Rip G. Rice, Incorporated
Ashton, Maryland 20861

Abstract: Since its adoption by the City of Nice, France in 1906 for disinfection of drinking water, the use of ozone as a water treatment agent has grown in importance. Early installations used ozone primarily for disinfection at the end of the treatment process. In modern times, however, the use of ozone as a chemical oxidant during early stages of water treatment have become prevalent. In light of current recognition of problems of organics in drinking water, modern water treatment facilities are learning how to combine ozone as a chemical oxidant with other water treatment steps, including ozone disinfection. The overall benefit of two- or three-stage ozonation to the improved quality of water becomes one of removing detrimental organic materials to minimize the amount of chlorine compound residual required for the distribution system.

EARLY HISTORY (Hill and Rice, 1982)

Although ozone was first discovered in 1840 by Schönbein during electrolysis experiments, early studies on its use as a germicide for treating potable water were not conducted until 1886, by de Meritens in France. Early pilot plant tests of ozone disinfection at potable water plants were conducted in Oudshoorn, the Netherlands (1893),

Paris, France (late 1890s), Lille, France (1898) and Paderborn and Wiesbaden, Germany (1902 and 1903, respectively).

In 1906 ozone disinfection was installed at the Bon Voyage water treatment plant in Nice, France. Because of the fact that ozone has been used to treat Nice city waters ever since, Nice has become known as the "birthplace of ozonation for potable water treatment". The Bon Voyage plant operated until 1970, during which time two additional potable water plants incorporating ozone were constructed as the city and its suburbs grew in size. In 1970 the new and completely modernized Super Rimiez plant (still incorporating ozone) was constructed on a mountaintop overlooking Nice, and the three older plants were shut down.

In 1916, Vosmaer listed a total of 49 European water treatment plants using ozone, 26 of which were located in France. As of 1977, at least 1,036 water treatment plants throughout the world were using ozone (Miller, *et al*., 1978), and about half of these were located in France. It is noteworthy that Switzerland, a country no larger in size than the state of New Jersey, had 150 operating water treatment plants using ozonation in 1977.

During 1977, there were 23 drinking water treatment plants in Canada using ozone, all except one in the Province of Quebec. At the same time, there were only 5 plants using ozone in the United States. Today there are 40 to 50 operational plants in Canada (several outside of Quebec Province), and 20 in the USA (Table I). Another five American plants are under construction.

PROPERTIES OF OZONE PERTINENT TO WATER TREATMENT

Generation

At normal temperatures and pressures encountered in drinking water treatment plants, ozone is a gas, which is prepared on-site. This is accomplished by passing a dry, oxygen-containing gas (usually air for water treatment plants) between two electrical plates separated by a ceramic dielectric and a narrow discharge gap. Under these conditions, part of the oxygen is converted to ozone:

$$3O_2 \rightleftharpoons 2O_3$$

It is important to recognize that the synthesis reaction produces rather low yields of ozone, usually 1 to 3% when air is the feed gas, and 2 to 6% when pure oxygen is the feed gas.

It is also noteworthy that the reaction is an equilibrium one, meaning that ozone also decomposes back to oxygen from which it is formed. Heat increases the rate of the reverse reaction. Therefore, because ozone generation produces relatively large quantities of heat as a by-product, efficient cooling of the ozone generator is important to maximize ozone yields.

Contacting

Ozone is only partially soluble in water, as a rule of thumb, about 13 times the solubility of oxygen. As a result, applying ozone to water involves gas/liquid contacting. This can be accomplished in a variety of ways, which involve tradeoffs between additional energy required by the various available contacting systems, versus contact chamber size and contact time.

For detailed discussions of the generation and contacting of ozone, the reader is referred to recent treatises by Carlins and Clark (1982) and Masschelein (1982), respectively.

Oxidizability

Table II shows the relative oxidation potentials of a number of chemicals. Of those materials readily available for water treatment, ozone is the most powerful, having an oxidation potential about 1.5 times higher than that of chlorine. This means that ozone will oxidize materials faster, will attain a level of disinfection faster (or with lower concentrations), and sometimes will oxidize materials that weaker oxidants will not oxidize. More detailed discussions of ozone oxidation of organic materials will be found in later sections of this paper.

Ozone Residual

In properly treated drinking water, a measureable residual of ozone can be attained readily in the treatment plant. In fact, attainment and maintainment of specific and reproducible levels of dissolved ozone residual is used as a means of controlling and automating ozone disinfection processes. However, this residual is not sufficiently stable in the distribution system to provide continuing bacteriostatic activity. Consequently, it is usual practice to follow ozone disinfection, at the end of the water treatment process, with small quantities of chlorine, chlorine dioxide, or chloramine, depending upon local conditions. More will be said on this subject in the next section.

When ozone disinfects or oxidizes, oxygen atoms are introduced into the chemical structures of the materials being oxidized. Free oxygen also is released into solution. No halogenated by-products of oxidation are produced upon ozonation. This is a significant difference between ozone, which acts as a disinfectant and chemical oxidant, versus

TABLE I. U.S. POTABLE WATER TREATMENT PLANTS USING OZONE (July 1984)

Location	Primary Purpose of Ozone	Startup Date	Av. Flow Rate mgd	m^3/day
IN OPERATION				
Whiting, IN	T & O	1940	4	15,142
Strasburg, PA	disinf.	1973	0.1	379
Grandin, ND	Fe & Mn	1978	0.05	189
Saratoga, WY	T & O	1978	3.5	13,249
Bay City, MI	T & O	1978	40	162,487
Monroe, MI	T & O	1979	18	73,119
Newport, DE	disinf.	1979	0.25	1,016
Newport, RI	THM precursors; T&O; color	1980	5	20,311
No. Tarrytown, NY	T & O	1980	1.2	4,875
Kennewick, WA	color;T&O	1980	3	12,187
Elizabeth City, NC	color	1981	5	20,311
Casper, WY	disinf.	1982	5	20,311
Ephrata Borough, PA	T & O	1982	0.145	549
New Ulm, MN	Fe & Mn	1982	2.6	10,562
South Bay, FL	color	1982	2.2	8,937
Rockwood, TN*	flocculation; THM precursors	1982	6	22,712
Potsdam, NY	color	1984	1	3,786
Beria, OH*	THM & T&O	1984	3.6	14,348
Belle Glade, FL*	color, THM precursors & algae	1984	6	22,712
Stillwater, OK	color	1984	5	18,927

(continued)

Location	Primary Purpose of Ozone	Startup Date	Av. Flow Rate mgd	m^3/day
UNDER CONSTRUCTION				
New York, NY	organics	1984	3 (pilot)	11,358
Hackensack, NJ	color, Fe & Mn; THM precursors	1986	100	378,540
Los Angeles, CA	microfloc-ulation & organics	1986	580	2,195,531
UNDER DESIGN				
Myrtle Beach, SC*	color; THM precursors	1987	30	121,865
Clinton, IL	Fe	1985	1.5	5,679
PILOT PLANT STUDIES				
Rocky Mount, NC	color;THMs			
Celina, OH	T&O; THM precursors			

* two-stage ozonation plants

TABLE II. RELATIVE OXIDATION POWER OF OXIDIZING SPECIES

Species	Oxidation Potential, volts	Relative Oxidation Power*
fluorine	3.06	2.25
hydroxyl radical	2.80	2.05
atomic oxygen	2.42	1.78
ozone	**2.07**	**1.52**
hydrogen peroxide	1.77	1.30
perhydroxyl radicals	1.70	1.25
hypochlorous acid	1.49	1.10
chlorine	1.36	1.00

* based on chlorine as reference point (= 1.00)

chlorine, which also acts as a disinfectant and chemical oxidant, but also as a chlorinating agent.

Solving the current problems of halogenated organics in water supplies involves trying to minimize the unwanted chlorinating reactions while maintaining the disinfecting and/or chemical oxidation capabilities of chlorine. One of chlorine's strongest advantages in water treatment is its ability to provide a stable bacteriostatic residual in the distribution system.

APPLICATIONS OF OZONE IN WATER TREATMENT (Rice *et al.*, 1981)

Bacterial Disinfection

When used for disinfection, ozonation normally is conducted at the end of the treatment process. This means that chemical addition, flocculation, sedimentation and filtration all have been accomplished prior to the water being ozonized. The efficiency of the pretreatment process(es) to remove organics and other ozone-demanding constituents will determine the amount of ozone required to attain disinfection. In turn, the amount of oxidizable organic materials remaining when ozonation is conducted will determine the amount and types of organic oxidation products which will be formed in the ozonized water.

Operational French water treatment plants treating river water by means of chemical addition, flocculation, sedimentation, and filtration prior to ozone disinfection in the suburbs of Paris, normally apply 2.0 to 4.0 mg/L of ozone for disinfection. This dosage can increase to over 6 mg/L during warm summer conditions.

Because ozone is a much stronger bactericide than chlorine, contact times necessary to attain the desired bacterial disinfection are shorter when using ozone. Under conditions which produce viral inactivation using ozone (maintenance of 0.4 mg/L of residual ozone over four minutes -- see next paragraph), bacterial disinfection will be attained simultaneously.

Viral Inactivation

Pioneering studies by French public health officials (Coin *et al.*, 1964, 1967) have established that poliomyelitis virus types I, II, and III can be inactivated above 99.9% when exposed to a dissolved ozone residual of 0.4 mg/L for a minimum period of time of four minutes. Subsequent to these studies, the city of Paris adopted the criterion of maintaining an 0.4 mg/L residual ozone over four minutes after the initial ozone demand of the water has been satisfied for viral inactivation. This standard subsequently was adopted throughout France when ozone is applied for disinfection. Ozone disinfection facilities in many other countries have been patterned after the French standard, and the World Health Organization recommends the same conditions for ozone disinfection. It should be understood, however, that only France has formally adopted these conditions as standard for ozone disinfection.

Post-Treatment After Ozone Disinfection

It was pointed out earlier that ozone does not provide a stable residual for the distribution system. Moreover, the addition of ozone to water simultaneously adds large volumes of air and/or oxygen to the water, because the concentrations of ozone in the gases exiting the ozone generator are only 1 to 3% when air is used as the feed gas, or 2 to 6% when pure oxygen is the feed gas. Thus the act of ozonation results in significant increases in the dissolved oxygen content of the water.

Finally, it should also be appreciated that when ozone oxidizes dissolved organic components of water supplies, those organics are seldom oxidized totally to CO_2 and water. Instead, the organic materials are **partially** oxidized, and the partial oxidation products contain more oxygen than did their precursors. As a general rule, partially oxidized organic compounds are more biodegradable than the initial materials. They have been started along their way to complete oxidation, ultimately to produce CO_2 and water.

Because of the above consequences of ozone disinfection, it is clear that without a stable ozone residual for the distribution system, along with a higher dissolved oxygen content and with a higher content of biodegradable, partially oxidized organic constituents, the distribution system now becomes a "garden of Eden" for bacterial regrowth. This type of regrowth was, in fact, observed in Swiss distribution systems when ozone disinfection was installed as the terminal treatment step. As soon as the causes of regrowth were recognized, however, addition of a small residual of chlorine dioxide was adopted (about 0.15 mg/L). This prevented bacterial regrowth from occurring in the distribution systems (Miller, et al., 1978).

There can be interactions between chlorine dioxide and ozone, however, and for this reason, many Swiss water treatment plants using ozone for disinfection, routinely filter ozonized water through granular activated carbon prior to addition of chlorine dioxide. GAC serves the dual purpose of destroying excess ozone in the water and adsorbing many of the remaining organics in the water, thus providing an even higher quality of drinking water (Rice et al., 1982).

Ozone Oxidation of Soluble Iron and Manganese

Ferrous iron is rapidly oxidized by ozone to ferric ions, which then hydrolyze, coagulate, and precipitate as the insoluble ferric hydroxide, according to the following equations:

$$Fe^{+2} + O_3 + H_2O \longrightarrow Fe^{+3} + O_2 + 2(OH)^-$$

$$Fe^{+3} + 3H_2O \longrightarrow Fe(OH)_3\downarrow + 3H^+$$

Similarly, divalent manganous ions are easily oxidized to the tetravalent manganic ions, which then hydrolyze rapidly forming the insoluble manganese dioxide:

$$Mn^{+2} + O_3 + H_2O \longrightarrow Mn^{+4} + O_2 + (OH)^-$$

$$Mn^{+4} + 4(OH)^- \longrightarrow Mn(OH)_4 \longrightarrow MnO_2\downarrow + 2H_2O$$

However, excessive ozonation of manganese compounds will continue the oxidation process, producing the soluble septavalent, permanganate ion, which is pink in color:

$$Mn^{+2} \text{ or } Mn^{+4} + 4O_3 \longrightarrow MnO_4^- + 4O_2$$

Permanganate is toxic, and must be prevented from entering water distribution networks, where it is slowly reduced to insoluble MnO_2, leading to buildup of manganate scales. To avoid this possibility, ozonation for oxidation of iron and/or manganese normally is conducted prior to filtration, at an early stage in the water treatment process. Holding the ozonized water 30 minutes allows traces of permanganate to oxidize some of the organics present, and be reduced to the insoluble tetravalent state:

$$MnO_4^- + \text{organics} \longrightarrow MnO_2\downarrow + \text{oxidized organics}$$

Decomplexing of Organically Bound Heavy Metals

When decaying vegetation is present in raw water supplies, there is a strong possibility that heavy metals, particularly manganese, can be present in the form of soluble organic complexes. Such complexes are not easily decomposed by chlorine, but ozonation easily liberates the heavy metal as a soluble cation. For this type of application, ozonation is conducted at an early stage in the treatment process, before filtration.

Oxidation of Inorganics (Cyanides, Sulfides, Nitrites, and Ammonia)

Toxic cyanide ions are readily oxidized by ozone to the much less toxic cyanate ion:

$$(CN)^- + O_3 \longrightarrow (CNO)^- + O_2$$

Cyanate ion slowly hydrolyzes to produce CO_2 and nitrogen:

$$2(CNO)^- + 2H_2O \longrightarrow 2CO_2 + N_2 + 4H^+$$

Sulfide ion is easily oxidized to sulfur, then to sulfite, then to sulfate:

$$S^{-2} + O_3 \longrightarrow S^o \longrightarrow SO_3^{-2} \longrightarrow SO_4^{-2}$$

The degree of oxidation obtained will depend upon the amount of ozone employed and the contact time.

Nitrite ion is oxidized very rapidly by ozone to nitrate ion:

$$NO_2^- + O_3 \longrightarrow NO_3^- + O_2$$

Oxidation of these inorganic species will occur at any stage of the treatment process at which ozone is applied. Since the products of ozonation all are soluble (except for elemental sulfur from sulfide oxidation), there is no particular advantage to be gained in applying ozonation at any specific point in the treatment process, as is the case for oxidation of organics and for disinfection.

Under pH conditions normally found in water treatment plants, ammonia does not react with ozone at any appreciable rate below pH of 9.0 (Singer & Zilli, 1975). Therefore, ammonia in the raw water supply will carry through the plant, unless it is converted to nitrate biologically (nitrification - see section on Ozone Pretreatment for Biological Processing). Levels of ammonia generally are undesirable in finished waters because ammonia combines with free chlorine to produce chloramines, which are less beneficial as disinfectants than is free residual chlorine. However, chloramines do exert sufficient bacteriostatic action to be useful in distribution systems for residual.

Oxidation and Removal of Organics

Color Removal - Usually the colors present in drinking water are derived from the decomposition of naturally occurring humic materials, and are caused by the presence of conjugated unsaturated moieties, called chromophores. Ozone is particularly reactive toward unsaturated groups, cleaving the carbon-carbon double bonds to produce ketones, aldehydes, or carboxylic acids, depending upon the other substituents on the carbon atoms affected, the amount of ozone applied, the ozone contact time involved, temperature, etc. As soon as the conjugation has been disrupted by ozone, the color disappears. However, this does not mean that all of the color-causing compound has been totally oxidized to CO_2 and water. Instead, the original organic material has been partially oxidized, and now contains a large number of polar, oxygen-containing groupings.

These polar groupings now can combine with polyvalent cations, such as aluminum or ferric ions, added as flocculating agents, and thus can be easily precipitated from the treated water. Therefore, when using ozone for color removal, it should be applied prior to chemical addition and followed by filtration.

Oxidation of Other Organics - Organic compounds which cause taste and odor problems (many phenols, metabolic products of algal metabolism), and some detergents and pesticides can be oxidized partially upon ozonation. In most cases, this amount of oxidation is sufficient to destroy the undesirable organoleptic properties imparted to the water by the particular contaminant. Total oxidation rarely occurs, but the more polar oxidation products now are more easily flocculated and removed from solution. Consequently, ozonation for these purposes also should be conducted before chemical treatment and filtration.

Algae Control - During seasonal climate changes and when proper nutrient balances are present in the raw waters, algae growths are promoted in the presence of sunlight. Large amounts of algae will clog filters in the treatment plant, requiring more frequent backwashing. Ozonation disrupts the metabolic processes of many types of algae by oxidizing key organic constituents.

In treatment plants already employing preozonation for other purposes, seasonal blooms of algae in the raw waters are handled merely by increasing the preozonation dosage for as long as the algae bloom lasts (usually only a few weeks). In plants employing ozone for disinfection, seasonal blooms of algae are controlled by adding a contacting basin to the early stages of water treatment and preozonizing at this point.

Turbidity Control and Microflocculation - When turbidity is caused by colloidal sized suspended solids, ozonation sometimes will change the chemical nature of the particle surfaces, allowing the particles to coagulate and settle. In such instances ozonation decreases the level of turbidity.

In other cases, such as when ozone disinfection is conducted on waters containing relatively large quantities of dissolved organics, the turbidity of the ozonized water sometimes will increase. This is because of the introduction of oxygen into the organic materials, producing carboxyl, carbonyl, and hydroxyl groups. These highly polar moieties now can combine with polyvalent cations also present, such as calcium, magnesium, iron, aluminum, etc., forming higher molecular weight structures, which now precipitate. This phenomenon of turbidity formation after ozonation has been termed "microflocculation" by Maier (1979).

If the level of dissolved organics is lowered prior to ozone disinfection, microflocculation is not observed. The Swiss practice of filtration following ozone disinfection guarantees removal of any turbidity resulting from microflocculation.

In 1986, the City of Los Angeles, California will start up what will be the largest drinking water treatment plant in the world to utilize ozone. Some 3,311 kg (7,300 lbs) per day of ozone will be generated from oxygen to treat 2.2 million m^3 per day (580 mgd) to lower turbidity to less than 0.3 unit. The ozonized and chemically treated (iron salts and cationic polymer) raw water will be coagulated, flocculated, then filtered through six-feet of anthracite coal at a rate of 13.5 gal/min/ft^2. This high filtration rate is made possible by employing ozone as a microflocculant, without which filtration rates of the coagulated and flocculated water could not exceed 9 gal/min/ft^2 (Stolarik, 1983).

<u>Ozone Pretreatment For Biological Processing</u> - Examples cited earlier pointed out that under ozonation conditions normally applied in treating drinking water, most oxidizable organic constituents are only partially oxidized. However ozone does introduce considerable oxygen into these compounds and renders them more biodegradable.

At the same time, ozonation increases the dissolved oxygen content of the water, thereby improving conditions for aerobic microbiological activity. Such biological activity can convert dissolved organic impurities to CO_2 and water, but also can convert ammonia to nitrate.

Post-ozonation biological treatment can be conducted in a variety of manners. Raw waters can be preozonized, then stored several days in a reservoir before entering the treatment plant proper. After entering the treatment plant, intake waters can be ozonized along with chemical addition, then filtered. Ozonation partially oxidizes dissolved organics, which then are more completely precipitated by means of the flocculating agents. In sand filters, some dissolved organic carbon is removed biologically. Nitrification also occurs in sand filters.

GAC filters/adsorbers placed after sand filter beds also contain considerable aerobic bioactivity. Those dissolved organic compounds which have not been oxidized, even by ozone, are adsorbed by the GAC. Other organics which have been partially oxidized during ozonation may or may not be adsorbed, but in any case also can be biodegraded by the microorganisms present.

Because of the increased biodegradation which occurs in the GAC adsorbers, the breakthrough time for the GAC is postponed. In European water treatment plants which are employing this "biological activated carbon" process, it has been shown that the useful lifetime of the GAC can be extended by factors up to six times, simply by incorporating ozone oxidation prior to filtration through sand/anthracite.

Extension of the GAC operating life results in cost savings which, in some instances, have been shown to pay for the costs of installing preozonation. This aspect will be discussed further later in this chapter under "costs".

Ozone Treatment of THMs or THM Precursors - When chlorine is added to water containing trihalomethane precursors, trihalomethanes (THMs) and other halogenated compounds are produced. Once formed, THMs are not oxidized, even by ozone. The more volatile THMs may volatilize partially, because of the aerating action which accompanies the application of ozone to water. Adsorption on GAC will remove THMs for a short time, but rapid breakthough is observed, requiring frequent reactivation of the GAC, which is a costly process.

It is axiomatic that once THMs are produced, chlorine has been wasted and THMs are costly to remove. Thus, the approach of the discerning water treatment engineer should be to modify his treatment process to minimize the concentrations of THM precursors before the addition of chlorine. This approach should be contrasted with that of making no changes in the treatment process, but installing treatment techniques to remove THMs after they are formed.

Ozonation has been shown to lower the concentrations of some THM precursors, but in several cases THM precursor levels actually increase upon ozonation (Rice, 1980). As a result, the use of ozone to oxidize THM precursors and reduce their concentrations should be tested on each water supply before this approach is adopted.

THM precursors can be partially oxidized by ozone, then flocculated with standard flocculating agents. Thus, preozonation of raw waters, followed by chemical addition, flocculation, sedimentation and filtration appears to be one effective approach to lower the concentrations of THM precursors, and therefore, the concentration of THMs ultimately produced. Since the water treatment objective is to minimize the amounts of halogenated compounds, then prechlorination must be abandoned in favor of precursor minimization techniques.

Summary

Figure 1 shows the "conventional" water treatment process, along with appropriate points of application of ozone for specific purposes. In this figure, the "conventional" water treatment process is defined as treatment of the raw water with chemicals (other than chlorine) for coagulation, flocculation, sedimentation, then filtration and disinfection.

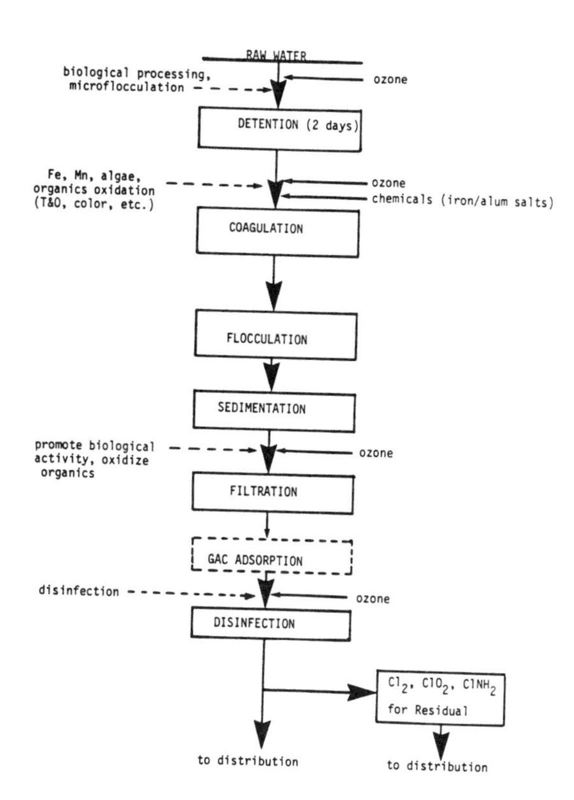

Figure 1. Points of ozonation during water treatment.

When ozone is used as an oxidant, it is applied during early stages of the treatment process. When used as a disinfectant, it is applied as the last treatment step prior to adding the appropriate chlorine compound for residual. At any point of application, ozone will both oxidize and disinfect. In early stages of treatment, when levels of oxidizable materials predominate over microorganisms, only partial disinfection may be achieved, unless large amounts of ozone are applied in relation to the amounts of oxidizable materials present.

Conversely, when ozone is used as a disinfectant after removal of most of the oxidizable constituents, complete disinfection can be attained with small amounts of ozone. Under these circumstances, only small amounts of organic oxidation products will be produced. Therefore, ozone should not be relied upon as a disinfectant in the early stages of treatment.

TWO-STAGE APPLICATIONS OF OZONE

The benefits of applying ozone at more than one point in the treatment process have been pioneered in modern French and German plants, and several plants in the Paris suburbs recently have installed <u>triple</u>-stage ozonation to improve the quality of water even further. Triple-stage ozonation will be discussed in a later section.

Rouen-la-Chapelle, France

One of the first plants to use double-stage ozonation is the la Chapelle plant at Rouen, France (Gomella and Versanne, 1977). Figure 2 is a schematic diagram of the water treatment process employed. Water is drawn from wells adjacent to the Seine River (well downstream of Paris). Preozonation is conducted to oxidize iron and manganese, as well as to partially oxidize dissolved organic constituents and to prepare the medium for nitrification and biological removal of organics.

Following ozonation, the water is sand filtered. Precipitated iron and manganese hydroxides are removed and most of the ammonia is converted to nitrate biologically. Next, the filtered water is passed through granular activated carbon beds where most of the organics are removed (by adsorption and biological degradation). Ozone disinfection, then post-chlorination for residual complete the treatment process.

Ozone disinfection is conducted in two contact chambers containing diffusion bubbling devices. These allow adequate ozone transfer efficiency to the water (90% and higher) without requiring additional energy beyond that necessary for the generation of ozone. In the first contact chamber, a level of 0.4 mg/L of residual ozone is attained during

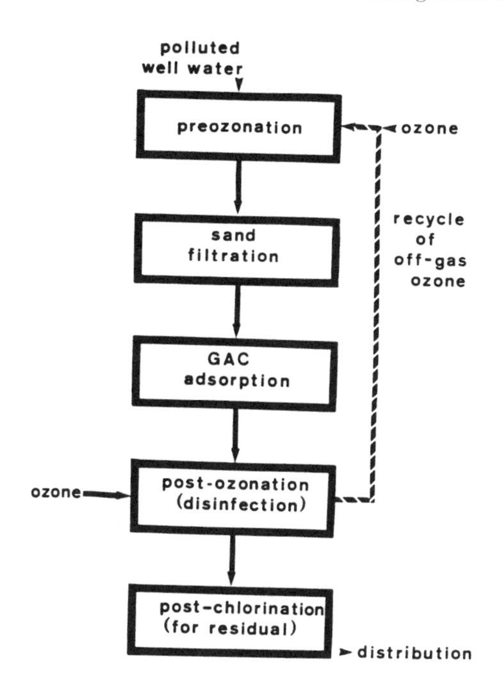

Figure 2. The Rouen-la-Chapelle process, Rouen, France.

a retention time of five to six minutes. In the second chamber, the 0.4 mg/L level of residual ozone is maintained during another five to six minutes. About two-thirds of the ozone used for disinfection is applied in the first contact chamber (to satisfy the initial ozone demand) and the balance is added in the second.

Because the transfer efficiency of ozone into water in the bubbler diffuser contactor is about 90%, the exhaust gases from the ozone disinfection contactor still contain a significant amount of ozone. If disinfection were the sole purpose of ozonation, these exhaust gases would be passed through an ozone destruction unit before discharge to the outside atmosphere. However, because of the second application of ozone at this plant (preozonation), the disinfection contactor exhaust gases instead are drawn into the preozonation contact chamber by means of a turbine, which also serves as the ozone contacting device and for rapid mixing.

This use of contactor exhaust gases serves two purposes: first, it supplies most of the ozone required for the first stage, and second, it utilizes most of the ozone which otherwise would have to be destroyed without realizing any benefit.

During the preozonation step there is no requirement to develop a specific residual of dissolved ozone for a specified period of time, because the objective is oxidation, not disinfection. Oxidation of iron and manganese and the phenols (which are the primary organics present) is nearly instantaneous. Partial oxidation of other organics, while not instantaneous, nevertheless occurs in the short contact time (two minutes at Rouen). Thus the use of a turbine contactor device, which requires energy for its operation, is cost-effective for the preozonation step. Not only does it provide violent agitation of ozone with the water, but it also creates a partial vacuum which draws the disinfection contactor exhaust gases into the preozonation contact chamber, thus eliminating the need for additional pumping.

In the event that more ozone is required for the preozonation step than is available in the disinfection contactor exhaust gases, the additional ozone can be provided directly from the ozone generators.

It is also worth noting that this two-stage ozonation system which includes GAC adsorption was installed at Rouen instead of a process involving prechlorination, followed by filtration, GAC adsorption (to remove chlorine and chlorinated organics) then ozone disinfection. This decision was based on considerations of costs; the process involving chlorine would have required significant space for long detention time chlorine contacting. In addition, the GAC used for dechlorination would have required frequent reactivation (less than six months) (Gomella and Versanne, 1977).

After three years of successful operation, it has been reported that the GAC had shown no signs of breakthough of organics sufficient to require its reactivation (Schulhof, 1979).

Mülheim, Federal Republic of Germany

One of the first water treatment plants to employ two-stage ozonation in the Federal Republic of Germany, was the Dohne plant in Mülheim. This plant treats Ruhr River water by the process shown schematically in Figure 3 (Sontheimer, et al., 1978), which has replaced the original process involving breakpoint chlorination of the raw water. By the new process, raw water is treated with flocculating agents and ozone under high speed agitation. This not only provides the desired degree of mixing, but also allows

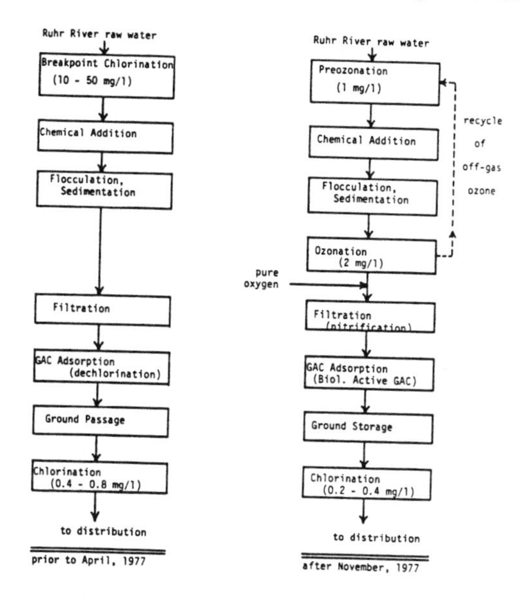

Figure 3. Comparison of the old (prechlorination) and new (two-stage ozonation) processes at the Dohne plant, Mülheim, Federal Republic of Germany.

effective contacting of ozone and partial oxidation and microflocculation of organic components.

After sedimentation, the supernatant water again is ozonized, this time in a two-chamber contactor with bubble diffusers using the French disinfection conditions similar to those employed at the Rouen-la-Chapelle plant. This treatment continues the oxidation of organic materials, making them all the more biodegradable.

Pure oxygen is added to the ozonized water to increase the dissolved oxygen level in order to maximize nitrification in the sand filters, and to promote aerobic biological activity in the GAC adsorbers.

As at the Rouen-la-Chapelle plant, exhaust gases from the second ozonation step are drawn into the preozonation stage for reuse/destruction of ozone.

Suburbs of Paris, France

Gerval (1978) reported experiences of the Syndicat des Communes de la Banlieue de Paris pour les Eaux with preozonation of raw waters to improve biological removal of dissolved organics and conversion of ammonia to nitrate. In the large plants of the Paris suburbs where ozone disinfection had been in use for many years with conventionally treated Seine, Oise and Marne river waters (including prechlorination), the levels of ozone normally required for disinfection historically averaged 2 to 4 mg/L. However, in summer, these values could rise to as high as 5 to 7 mg/L, primarily because of the presence of higher concentrations of ozone-demanding organics and increased microorganism activity.

When ozone was first installed at these plants, ammonia was removed by breakpoint chlorination of the raw water. However, with increasing levels of ammonia in the raw water, and with the increased production of halogenated organics during prechlorination, the treatment process had to be modified. The point of first introduction of chlorine has been moved to after the ozone disinfection step, and ammoniacal nitrogen now is being removed biologically.

Installations for biological removal of ammoniacal nitrogen at these plants consist of raw water reservoirs or detention tanks in which ozone-treated raw waters are stored approximately two days before entering the plant proper. The amount of ozone required for preozonation is small (less than 1.25 mg/L) and the contact time required is only two minutes.

Table III shows the effects of two-day reservoir storage with and without preozonation under the above conditions upon certain of the raw water quality parameters. Levels of detergents, organic carbon, COD and ammoniacal nitrogen are significantly lowered upon two-day storage preceeded by preozonation.

TABLE III. EFFECTS OF PREOZONATION BEFORE STORAGE (1.25 mg/L ozone dosage)

	detergents (mg/L)	organic carbon (mg/L)	COD (mg/L of oxygen)	NH_4^+ (mg/L)
raw water	0.16	8.3	19	6.2
2-day storage (no preozonation)	0.13	8	11	4
2-day storage (with preozonation)	0.08	7	6	3.2

Of even more significance was the fact that the preozonation dosage of 1.25 mg/L lowered the post-ozonation disinfection dosage from an average of 4.2 mg/L to 2.1 mg/L. Thus the total amount of ozone required at this plant has been kept the same (or even slightly reduced), while the only investment cost involved was the installation of a preozonation contactor system and the associated piping.

Further benefits resulting from this process modification are:

- a reduction in quantities of pretreatment chemicals formerly required,
- an approximately 50% increase in filtration cycle time,
- oxygenation of the water, which improves the elimination of nitrogen even further.

When the filter inlet ammoniacal nitrogen level is not greater than 1 mg/L, the outlet level does not exceed 0.1 mg/L.

Because of these benefits of preozonation, it is now possible for the plant to use less chlorine downstream for post-disinfection residual, which, in turn, lowers THM contents to below 10 mg/L in the distribution systems.

When the preozonation system was optimized at this Paris suburbs plant, it was found that the usual 4 mg/L ozone disinfection dosage (when this was the sole ozonation step) was reduced to 1.2 mg/L by applying an optimized preozonation dosage of 0.6 mg/L, an actual savings in total ozone requirement.

Since installation of two-stage ozonation in the Rouen-la-Chapelle plant in 1976 and at the Dohne plant in 1977, many other European water treatment plants have installed similar treatment processes.

Belle Glade, Florida

In the United States, the first two-stage ozonation water treatment plant is under construction at Belle Glade, Florida, and is scheduled to start operating in 1984 (Wagner and Elefritz, 1983; Elefritz *et al.*, 1984). Raw water for this plant is taken from Lake Okeechobee, which has very high contents of humic color and other naturally occurring organic materials, turbidity and biological growths. Normally the lake waters exhibit COD levels in excess of 60 mg/L with TOC levels about 30 mg/L. After storms, these levels have risen to as high as 100 and 75 mg/L, respectively, with color levels peaking at 500 color units.

Historically, this raw water has been prechlorinated for algae control and partial color removal, followed by lime softening and treatment with alum, synthetic polymers, and recycled lime sludge, then recarbonation prior to filtration and chlorine disinfection. By this process, the Belle Glade plant normally produces total THM levels of 600 to 900 microg/L (ppb). Moving the point of chlorination was not effective, in that serious algae blooms then were encountered within the treatment plant. The respiratory products of the algae and of the saprophytic bacteria were not removed by the conventional treatment process, without prechlorination, and taste, odor, and additional THM problems resulted upon postchlorination.

Figure 4 shows a schematic of the new two-stage ozonation process which was developed for treating the Lake Okeechobee water at Belle Glade. Raw water is preozonized to modify the structure of the THM precursors, making them more susceptible to physical removal by means of added lime and polyelectrolytes after flash mixing. After clarification, the water is recarbonated for pH control, then post-ozonized to polish remaining organics, filtered to remove any remaining apparent color and filterable organic material, then post-chlorinated to develop a free residual of chlorine for the distribution system.

This process was demonstrated over a 12-week period in a 5-gal/min pilot plant unit. During this period of time, terminal THM concentrations were maintained below 100 microg/L (ppb) consistently. Using free chlorine disinfection, the THM content could be maintained as low as 25 ppb consistently. With combined chlorine residual, the THM content could be adjusted to even lower levels. During the same period of time, the conventional process generated 600 to 1,600 ppb of THMs in the main plant.

By tracing the TOC through the pilot plant operation, it was shown that nearly 80% of the organics removed were carried out physically during the clarification and filtration steps.

The total amount of ozone to be applied in the plant is about 6 mg/L, half of this being added during preozonation, the other half during post-ozonation. As a result of improved removal of organics (with control of algae and color as additional benefits), the post-chlorination disinfection dosage now will be about 2 mg/L. This dosage is sufficient to provide a stable free chlorine residual in the Belle Glade distribution system and to provide the very much lowered THM levels.

Wagner and Elefritz (1983) recommend that before initiating ozone testing, the plant process steps of mixing, clarification and filtration first should be adjusted to maximize efficiency(ies). At the Belle Glade plant, this included retrofitting clarifiers with vertical flow tube settlers to improve overall efficiency and capacity.

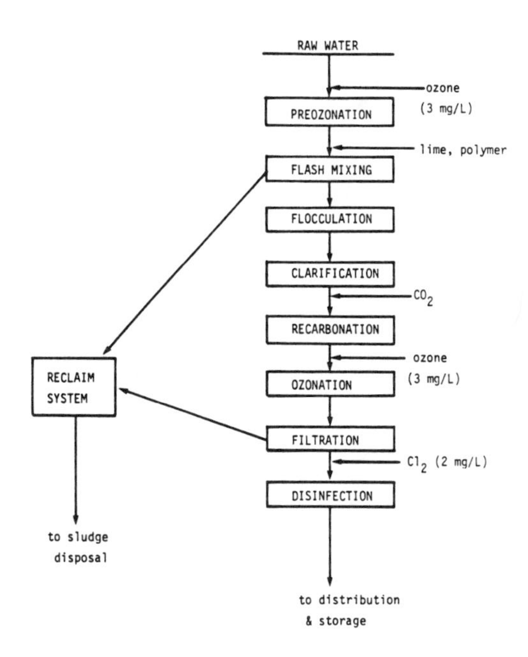

Figure 4. The Belle Glade, Florida, two-stage ozone treatment system (Wagner & Elefritz, 1983).

TRIPLE-STAGE OZONATION

Following successful experiences with the promotion of biological oxidation of organics and nitrification of ammoniacal nitrogen by use of preozonation, triple stage ozonation was designed into three large water treatment plants in the suburbs of Paris, France during the period 1980-1981 (Schulhof, 1980). Recently, operational experiences with this new process have been described for the plant at Méry-sur-Oise (Rapinat, 1982). The three-stage ozonation process installed and operating at this plant is shown in Figure 5.

Raw water is preozonized using 0.25 mg/L ozone dosage, then allowed to stand several days in a storage basin. During this time settling occurs, as well as some biological decomposition of dissolved organic materials, and some nitrification. Following chemical treatment, flocculation, sedimentation, and sand filtration, ozone again is applied prior to GAC adsorption, and again a third time for disinfection after GAC adsorption. Chlorine for residual is added as the water leaves the treatment plant by means of sodium hypochlorite solution.

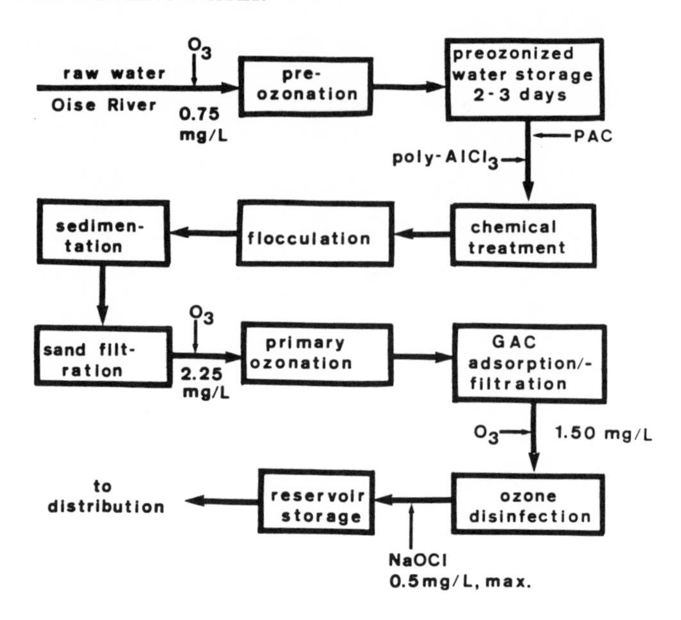

Figure 5. Triple-stage ozonation at Méry-sur-Oise (Paris, France) water treatment plant

The original treatment process at Méry-sur-Oise (breakpoint chlorination, flocculation, sand filtration, ozone disinfection and postchlorination) removed 66% of the dissolved organics from the Oise river water. By the new process, involving three-stage ozonation, biological treatment, and elimination of the prechlorination treatment, 80% of the dissolved organics are removed and the product water contains no significant levels of trihalomethanes. Table IV compares the effectiveness of the two processes in removing organic materials.

In the original process, the dosage of ozone required for disinfection averaged 3.5 mg/L. The new, triple-stage ozonation process requires a total of 4.5 mg/L. Total costs for the new process, including regeneration of the GAC every two years, are only 6% higher than those of the older, single-stage ozonation process. Relative costs for the two treatment processes are compared in Figure 6.

TABLE IV. ORGANICS REMOVAL AT MERY-SUR-OISE PLANT BY CHEMICAL AND BIOLOGICAL TREATMENT PROCESSES

after process step	percent removal by	
	chemical trtmt.	biological trtmt.
preozonation & storage	---	30%
flocculation & settling	---	45%
breakpoint chlorination & flocculation	50%	---
sand filtration	58%	60%
ozonation & post-Cl_2	66%	---
ozonation	---	68%
GAC adsorption	---	74%
post-ozonation	---	78%
post-chlorination	---	80%

COSTS FOR OZONE TREATMENT OF DRINKING WATERS

In this section, costs are expressed in U.S. dollars and volumes are in cubic meters and U.S. gallons.

In surveying European and Canadian water treatment plants using ozone, Miller <u>et al</u>. (1978) found that capital costs for ozonation systems can range from a low of $1,100/kg ($500/lb) of ozone generation capacity per day for large systems to a high of $8,800/kg ($4,000/lb) per day for relatively small systems. Added costs are housing for the generating system, which can range from 20 to 33% of equipment costs, and contacting system costs.

Operating costs for ozonation systems depend upon energy demand, amortization rates, amortization period, and local energy costs. In surveying European drinking water treatment plants, Miller <u>et al</u>. (1978) found these costs to range from $0.46¢/$m^3$ to 1.05¢/m^3 (1.75 to 4¢/1,000 gal) of water treated.

Lepage (1981) has described the 68,137 m^3/day (8-mgd) water treatment plant at Monroe, Michigan, which has been ozonizing for taste and odor control since early 1979. Two ozone generators, each capable of producing 3 kg/day (225 lbs/day) of ozone from dry air are in use. Total capital cost of the ozone generation and contacting system equates to $2,310/kg ($1,050/lb) of ozone generated.

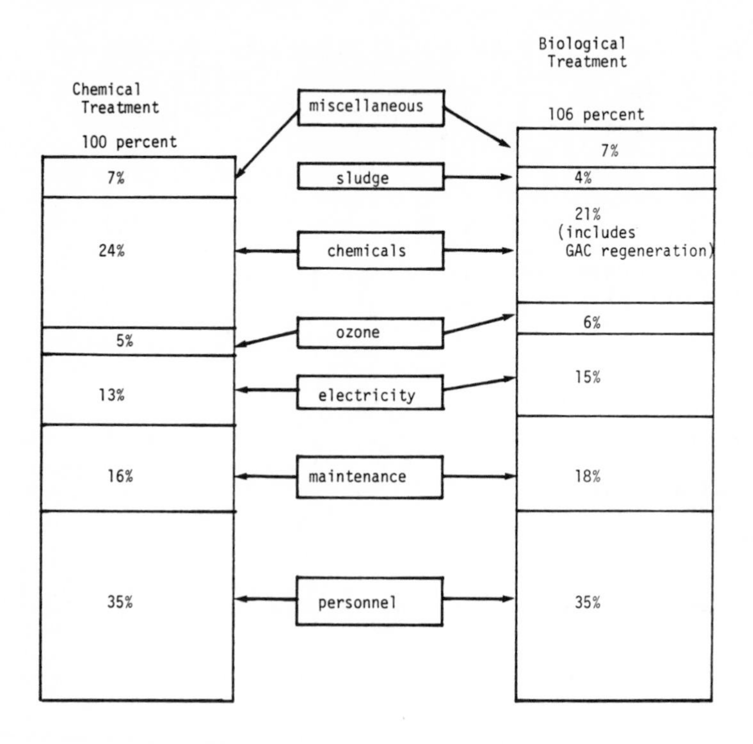

Figure 6. Comparative costs of chemical and biological treatment at Mery-sur-Oise (France) water treatment plant (Rapinat, 1982)

At an average applied ozone dosage of 1.63 mg/L to a 30,283 m^3/day (8-mgd) average flow rate and with an average ozone transfer efficiency in the contactor system of more than 97%, the total operating costs of air preparation, ozone generation, contacting, and reinjection of contactor exhaust gases have averaged 0.168¢/m^3 (0.636¢/1,000 gal) at Monroe.

It should also be noted that each of the two ozone generators installed at Monroe <u>can</u> handle the ozone demand of the system by itself if operated at its full production output. However, it is much more cost-effective to use both generators operated at less than full production capacity. The situation is analogous to operating an automobile at 100 miles/hr versus 50 miles/hr. At the slower speed, the energy demand is much lower.

When one ozone generator is taken off-line for preventive maintenance, the second generator is turned up to provide the ozone required for the few days involved.

At Bay City, Michigan, a 151,416 m^3/day (40 mgd) drinking water treatment plant has been operating since early 1980 for taste and odor control at an average flow rate of 45,425 m^3/day (12 mgd). Four ozone generators are capable of producing 250 lbs/day (113.4 kg/day) each, or a total of1,000 lbs (453.5 kg)/day of ozone. Capital costs for installing these ozone generators were $660,800, which equates to $661/lb of ozone generation capacity. Operating costs for an average applied ozone dosage of 1.5 mg/L were reported to be 0.118¢/m^3 (0.448¢/1,000 gal) during the first six months of operation (Croy, 1980).

At New Ulm, Minnesota, 9,811 m^3/day (2.59 mgd) of well waters are treated with an ozone dosage of 2.1 mg/L to oxidize iron and manganese. The capital cost for the ozonation system capable of generating up to 34 kg/day (75 lbs/day) from dry air was $480,000 ($6,400/lb of ozone generation capacity; contacting costs are probably included in this figure) in 1982 dollars (Kirk, 1984).

In Potsdam, New York, a new 9,842 m^3/day (2.6 mgd) water treatment plant went on-stream in the summer of 1983 (Simcoe, 1983). Ozone is used for disinfection of river water prior to chlorination. In pilot plant studies, the THM maximum formation potential was lowered from 353 microg/L to 116 microg/L with applied ozone dosages of 2.5 mg/L. A single ozone generator produces 150 lbs (68 kg)/day of ozone from dry air.

Costs for the ozone generation equipment, including instrumentation and controls, is $227,600 ($1,517/lb of generation capacity). Ozone contactor chambers cost $194,400, of which $73,500 is for reinforced concrete structures, $55,000 for stainless steel covers, and $65,900 for stainless steel wall pipes, access manways, and drain piping.

Estimated annual operating costs for ozonation at this facility are estimated to be $20,000, consisting of $7,200 for ozone generation and thermal destruction of contactor exhaust gases, $4,100 for plant heating and ventilating, and $8,700 for equipment maintenance. These costs equate to 0.021¢/1,000 gal.

The City of Los Angeles conducted extended cost analyses on alternative treatment processes for controlling turbidity, taste and odor, reduction in levels of trihalomethanes and for assuring bacterial and viral disinfection (Stolarik, 1983). Their final decision came down to ozonation versus chlorination, and their determinant cost figures with regard to the two processes are shown in Table V. By integrating ozonation into the filter plant design, a lower ultimate capital cost project results.

TABLE V. EFFECT OF OZONE VS. CHLORINE SYSTEMS COSTS ON TOTAL PROJECT COST* AT LOS ANGELES (Stolarik, 1983)

Capital Cost Item	Ozone	Chlorine
Process Equipment	$ 5,661,000	$ 166,000
Related Process Structures and Appurtenances	5,461,700	---
Additional Filtering Capacity (related costs)	---	13,070,330
Related Electrical	1,160,000	782,000
Additional Backwash Capacity	---	1,260,500
Miscellaneous	1,094,300	40,000
Subtotal	$13,790,800	$15,318,830
[Present construction provision for future facilities (O_3) should Cl_2 not achieve taste & odor control, THM standard]		
(including sales tax) TOTAL	$13,790,800	$18,532,830
Operation and Maintenance Item		
Chlorine	$ ---	$ 162,400
Additional Chemical Coagulant	---	314,000
Ozone Dielectric Maintenance	127,000	-------
Additional Power	498,000	42,100
TOTAL	$ 625,000	$ 518,500

*Note: The costs reflect the difference between the alternatives as they affect the total project cost of the filtration plant, and do not include the costs for features which are the same.

Annual operation and maintenance costs for ozone are higher than for chlorine, but the 50-year present worth (using a 9% discount rate) of the difference ($1,167,300) was less than the savings in plant construction cost due to the use of ozone.

An earlier present worth comparison between ozone and chlorine dioxide showed that the present worth cost of chlorine dioxide for this plant would be twice that of ozone (Stolarik, 1984).

At Belle Glade, Florida, capital costs for equipment to generate 500 lbs (227 kg)/day of ozone, and to install the entire ozonation system, including contact reactors, piping, and a building to house the generators was $670,000 (Wagner and Elefritz, 1983). This equates to $1,340/lb of ozone generation capacity. Over 20 years at 12% compounded annual interest, this capital amounts to 3¢/1,000 gal of water to be treated.

Elefritz, et al. (1984) reported that the complete turnkey cost of the ozonation system was $530,000 for the 500 lbs/day system. This equates to $1,060/lb of ozone generation capacity.

Table VI summarizes the estimated chemical requirements for the conventional and new dual-stage ozonation process at Belle Glade. It can be seen that costs for lime, polymer, and carbon dioxide are the same in both processes, but that alum is eliminated (saving $99,621/yr), and chlorine costs are lowered from $58,619/yr to $7,032/yr. Ozone costs add $101,000/yr; thus the new process saves about $50,000/yr in chemical costs. Stated another way, the chemical costs for the two processes are:

Conventional Process	14.81¢/1,000 gal
2-Stage Ozone Process	12.84¢/1,000 gal

The nearly 2¢/1,000 gal difference represents a 13% savings in chemicals costs.

Table VII summarizes the capital and operating cost data for ozone at the six U.S. water treatment plants discussed above. With the exception of the plant at New Ulm, Minnesota, the capital costs range from $661 to $1,517 per pound of ozone generation capacity. Operating costs for ozonation range from 0.021¢/1,000 gal at Potsdam, New York to 3.95¢/1,000 gal at Belle Glade, Florida.

ORGANIC OXIDATION PRODUCTS FROM OZONATION

Ozone oxidation of organic materials produces many intermediate oxidation products, as the carbonaceous materials progress along their oxidation pathways ultimately to produce CO_2 and water. As a general rule, these oxidized intermediates all contain more oxygen than did their precursors. Also as a general rule, the more oxygen organic compounds contain, the more water soluble and the more biodegradable they are. This is why ozone oxidation of organic materials promotes biodegradation.

TABLE VI. ESTIMATED CHEMICAL REQUIREMENTS AT BELLE GLADE, FLORIDA (Elefritz et al., 1984)

CONVENTIONAL PROCESS @ 7 MGD

CHEMICAL	DOSAGE (mg/l)	UNIT COST ($/kg)	DEMAND (kg/DAY)	COST ($/DAY)	COST ($/YEAR)
LIME	200	0.0159	5307.27	408.66	149,161.00
ALUM	0.55	0.0386	1459.55	272.94	99,621.28
CARBON DIOXIDE	20	0.0636	530.73	163.46	59,664.36
POLYMER	0.20	1.2045	5.32	31.00	11,316.83
CHLORINE	25	0.0500	663.64	160.60	58,619.00
				$1,036.66	$378,382.36

AVG-0.0391$/$m^3$(14.81¢/1000GALS)

MODIFIED OZONE PROCESS @ 7 MGD

CHEMICAL	DOSAGE (mg/l)	UNIT COST ($/kg)	DEMAND (kg/DAY)	COST ($/DAY)	COST ($/YEAR)
LIME	200	0.0159	5307.27	408.66	149,161.00
POLYMER	0.20	1.2045	5.32	31.00	11,316.83
CARBON DIOXIDE	20	0.0636	530.73	163.46	59,664.36
* OZONE	6	0.3591	159.09	276.72	101,003.24
CHLORINE	3	0.0500	79.61	19.27	7,031.87
				$899.11	$328,177.30

AVG-0.0339$/$m^3$(12.84¢/1000GALS)

* OZONE @ 12 KILOWATT HOURS/LBS @ 6.5¢ PER KWHR

TABLE VII. SUMMARY OF OZONE CAPITAL AND OPERATING COSTS AT OPERATIONAL U.S. WATER TREATMENT PLANTS

plant (yr started)	size m^3 (mgd) per day	amt. of O_3 generated kg (lbs) per day	capital cost $/lb capacity	operating cost ¢/1,000 gal trtd.
Monroe, MI (1979)	68,137(18)	113(225)	$1,050	0.636¢
Bay City, MI (1978)	151,416(40)	454(1,000)	$ 661	0.448¢
New Ulm, MN (1982)	9,811(2.6)	341(75)	$6,400	not given
Potsdam, NY (1983)	9,842(2.6)	68(150)	$1,517	0.021¢
Los Angeles, CA (1986)	2.2MM(580)	3,311(7,300)	$ 716	0.295¢
Belle Glade, FL (1984)	22,712(6)	227(500)	$1,340* $1,060	3.95¢

* including costs of two-stage contacting + housing

Improved biodegradability also implies less toxicity of the oxidation products, but this is not always the case, particularly when aromatic rings are involved. With phenol, for example, initial oxidation takes place without cleaving the ring, and it should be assumed that the polyhydroxyaromatic intermediate oxidation products are as toxic as is phenol. Once the aromatic ring is ruptured, however, the aliphatic, carboxyl and carbonyl-containing products become less toxic, and more biodegradable.

Figure 7 shows many of the oxidative pathways which phenol undergoes when it is oxidized (Miller, et al., 1978). Included in this figure are dichlorinated phenols and alkylphenols, which are shown to produce the same oxidation products as phenol after ring rupture. The carbon-chlorine bonds are broken at some point during the oxidation process, generating chloride ions. Although the reactions in Figure 7 are shown as being caused by ozone oxidation, it should be appreciated that many of these intermediate oxidation products also can be produced by other strong oxidizing agents, including chlorine dioxide, permanganate, hydrogen peroxide, and chlorine itself.

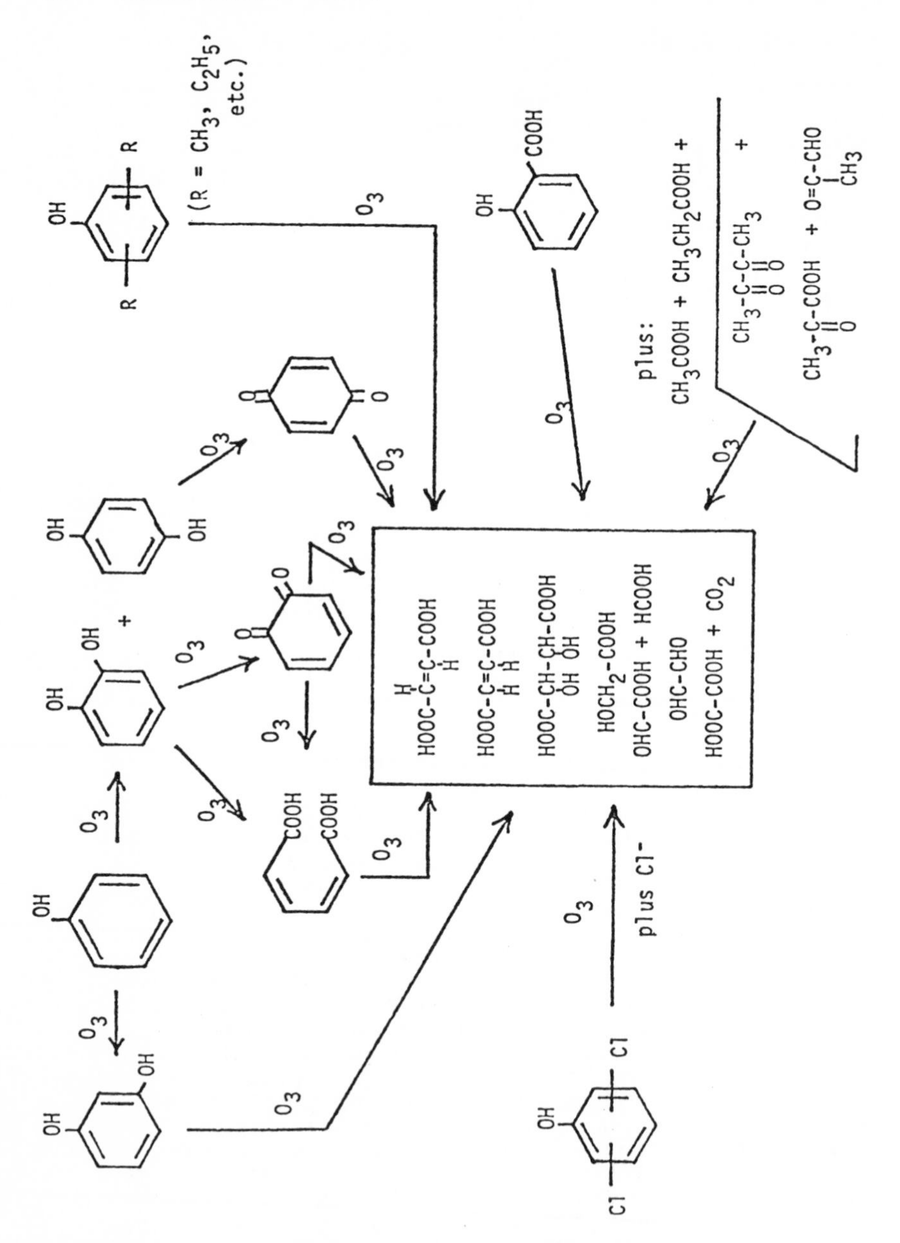

Figure 7. Reactions of ozone with phenols (Miller *et al.*, 1978).

Certain pesticides, such as parathion and malathion form their respective oxons upon initial oxidation. Further oxidation (using more oxidant and/or longer contact times) will convert these intermediates to other products, as shown in Figure 8. However, it should be appreciated that the oxons are more toxic than are the thions (Martin and LaPlanche, 1977).

The strong insecticide Heptachlor upon ozonation produces the more toxic Heptachlorepoxide, which is stable to further treatment with ozone under conditions normally employed in drinking water treatment plants (Hofmann and Eichelsdörfer, 1971).

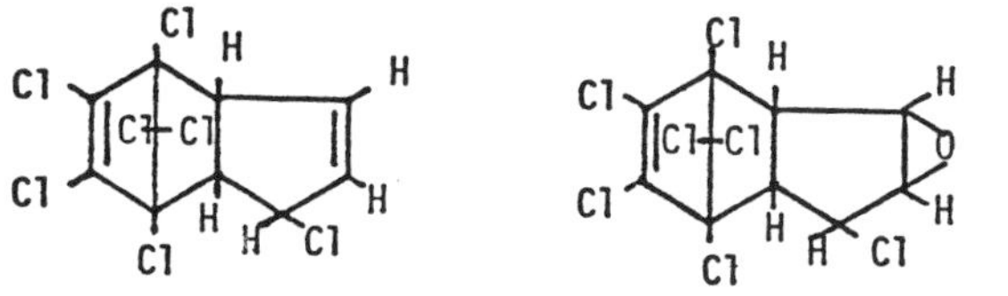

heptachlor

heptachlorepoxide

These pesticide examples lead to several significant conclusions:

1) To the maximum extent possible, the water treatment specialist should know what oxidizable materials are in his raw water, *and* what products are formed upon oxidation,

2) If ozone, the most powerful oxidizing agent employed in drinking water treatment, will not totally oxidize some organic compounds to CO_2 and water, *neither will weaker oxidizing agents*,

3) The application of *any* oxidizing agent to waters containing high concentrations of dissolved organic materials will produce organic oxidation products.

4) The proper water treatment strategy should be to remove organic materials to the maximum extent possible by physical and chemical techniques before any oxidant is added.

In reviewing the published literature on the oxidation products of organic materials produced by ozone or by chlorine dioxide, Miller *et al*. (1978) concluded that, as a general rule, *all* oxidants used for treatment of drinking water (chlorine, chlorine dioxide, permanganate, and ozone) can produce similar, and in some cases identical, non-halogenated oxidation products. Each oxidant will produce other organic oxidation products that are unique

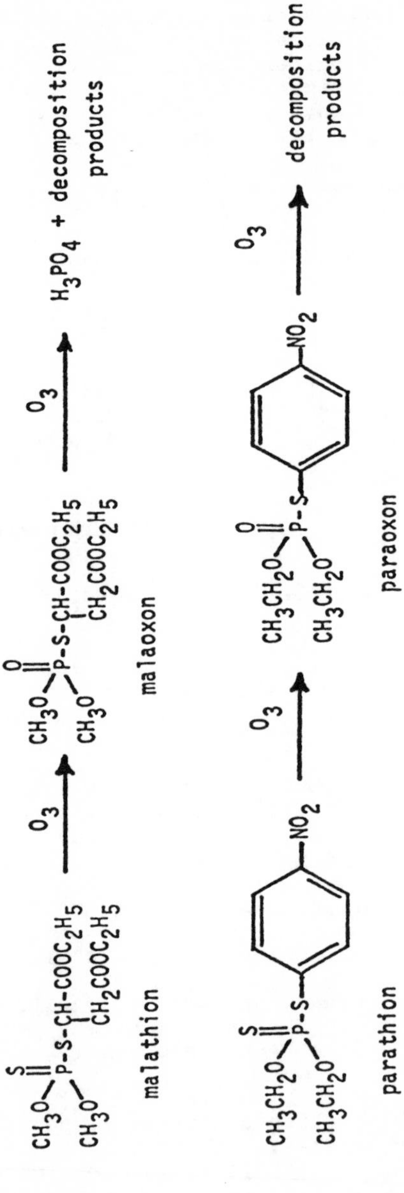

Figure 8. Ozonation of Malathion and Parathion (Martin & Laplanche, 1977).

to its chemistry, such as halogenated organics when using chlorine, and from chlorine dioxide under certain conditions.

This conclusion is supported by the work of Carlson and Caple (1977) who studied the oxidation of a variety of organic compounds (alpha-terpineol, oleic acid, abietic acid and cholesterol) with chlorine and with ozone in aqueous solutions. Many of the oxidation products isolated and identified were identical, regardless of which oxidant was employed.

CONCLUSIONS

1) The use of ozone for treating drinking water is a common practice, primarily in Europe where its use was pioneered in the early 1900s. Today there are at least 1,300 ozone water treatment plants throughout the world, 40 to 50 plants in Canada and 20 in the United States, successfully employing ozone for one or more purposes.

2) Ozone serves both as a disinfectant and as a chemical oxidant, and both processes will be operative regardless of where ozone is added during the treatment process.

3) When used as a disinfectant, ozone is applied at or near the end of the treatment process; when used as an oxidant, ozone normally is applied at an early stage.

4) Because ozone does not provide a stable residual for distribution systems, a chlorine compound (chlorine, chlorine dioxide or chloramines) normally is added for this purpose, after ozone disinfection.

5) The ability of preozonation to partially oxidize organic materials in raw waters leads to their more efficient removal by subsequent flocculation and filtration. In many cases, preozonation saves chemical costs. At the new Los Angeles, CA plant, preozonation with chemical treatment will allow a 50% faster filtration rate.

6) Most organic materials cannot be oxidized completely to CO_2 and water under ozonation conditions normally employed in water treatment plants. Partial oxidation products generally are more biodegradable than the original compounds. Thus ozonation can be used to promote biological activity, which when properly designed and operated within the water treatment plant, can lower the total organic carbon and ammonia contents of the treated water.

7) Plants already using ozone for disinfection can incorporate preozonation at the incremental cost of adding a preozonation contacting system. Ozone-containing exhaust gases from the disinfection contactor generally provide the requisite amount of ozone for the preozonation step.

8) Preozonation has replaced breakpoint prechlorination at many European water treatment plants, thus moving the point of chlorine addition until much later in the treatment process, after most of the organics have been removed. This results in much lower quantities of halogenated organics being produced.

9) Installing ozonation prior to GAC adsorption can prolong the useful life of the GAC up to a factor of six times. In many instances these savings in GAC reactivation can pay for the costs of preozonation.

10) The most modern French water treatment plants now are employing triple-stage ozonation (preozonation, ozonation prior to GAC adsorption, and ozone disinfection) to produce high quality drinking waters. This is being accomplished with less than 30% more ozone than that originally required for disinfection alone.

11) Although the capital costs of installing ozonation equipment are relatively high compared with other oxidants, operating costs are low. Capital costs for ozone generation in some of the newer U.S. water treatment plants range from $661 to $1,500 per pound of daily ozone generation capacity. Operating costs at the same plants range from 0.02¢ to 3.95¢ per 1,000 U.S. gallons treated.

12) Whenever <u>any</u> chemical oxidant is added to waters containing high concentrations of dissolved organic materials, oxidation products will be formed. Although each chemical oxidant produces organic oxidation products specific to its oxidation chemistry, many of the same products are formed by all chemical oxidants. Minimizing the formation of organic oxidation products requires maximizing the removal of organics by physical and chemical techniques before oxidants are added.

REFERENCES

Carlins, J.J., and R.G. Clark, 1982, "Ozone Generation by Corona Discharge", in Handbook of Ozone Technology and Applications, R.G. Rice & A. Netzer, Editors (Stoneham, MA, Ann Arbor Science Publishers, Inc.), p. 41-76.

Carlson, R.M., and R. Caple, 1977, "Chemical/Biological Implications of Using Chlorine and Ozone for Disinfection", EPA Report No. 600/3-77-066. Natl. Tech. Info. Service Report No. PB 270,694.

Coin, L., C. Hannoun, and C. Gomella, 1964, "Inactivation of Poliomyelitis Virus by Ozone in the Presence of Water", la Presse Médicale 72(37):2153.

Coin, L., C. Gomella, C. Hannoun, and J.C. Trimoreau, 1967, "Inactivation of Poliomyelitis Virus in Water", la Presse Médicale 75(38):1883.

Croy, R.S., 1980, "Bay City Plant is Newest, Largest in U.S. Using Ozone for Drinking Water Treatment", Part I: OZONEws 8(6)(June); Part II: OZONEws 8(7)(July). Intl. Ozone Assoc., Norwalk, CT.

Elefritz, R.A., D.W. Porter, and S.F. Morris, 1984, "The Application of Ozone in Softening Processes for Cost-Effective THM Control: Two Case Histories", presented at Seminar on Strategies for the Control of Trihalomethanes, Am. Water Works Assoc. Southeast Annual Conf., Jekyll Island, GA (April 29, 1984).

Gerval, R., 1978, "Ozonation of Raw Water Before Its Treatment in Potable Water Treatment Plants", l'Eau et l'Industrie 29:44-47.

Gomella, C., and D. Versanne, 1977, "Role of Ozone in the Bacterial Nitrification of Ammoniacal Nitrogen: Case of the la Chapelle Plant at St.-Etienne Rouvray (Seine Maritime)", Techniques Sanitaires et Municipales, p. 78-81.

Hill, A.G., and R.G. Rice, 1982, "Historical Background, Properties, and Applications of Ozone", in Handbook of Ozone Technology and Applications, R.G. Rice & A. Netzer, Editors (Stoneham, MA, Ann Arbor Science Publications, Inc.), p. 1-40.

Hofmann, J., and D. Eichelsdörfer, 1971, "On the Action of Ozone on Pesticides Containing Chlorohydrocarbon Groups in Water", Vom Wasser 38:197.

Kirk, J.T., 1984, "Ozonation Reduces Manganese Concentrations", Public Works 115(1):40-42.

Larocque, R.L., 1984, Hankin Environmental Systems, Toronto, Canada, Private Communication.

LePage, W.L., 1981, "The Anatomy of an Ozone Plant", J. Am. Water Works Assoc. 73(2):105-111.

Maier, D., 1979, "Microflocculation With Ozone", in Oxidation Techniques in Drinking Water Treatment, W. Kühn & H. Sontheimer, Editors, EPA Report No. 570/9-79-020, p. 394-417.

Martin, G., and A. LaPlanche, 1977, "Action of Ozone on Organophosphorus Compounds", presented at Intl. Ozone Assoc. Third Ozone World Congress, Paris, France, May.

Masschelein, W.J., 1982, "Contacting of Ozone With Water and Contactor Off-Gas Treatment", in Handbook of Ozone Technology and Applications, R.G. Rice & A. Netzer, Editors (Stoneham, MA, Ann Arbor Science Publishers, Inc.), p. 143-226.

Miller, G.W., R.G. Rice, C.M. Robson, R.L. Scullin, W. Kühn, and H. Wolf, 1978, "An Assessment of Ozone and Chlorine Dioxide Technologies for Treatment of Municipal Water Supplies", EPA Report No. 600/2-78-147.

Rapinat, M., 1982, "Recent Developments in Water Treatment in France", J. Am. Water Works Assoc. 74(12):610-617.

Rice, R.G., 1980, "The Use of Ozone to Control Trihalomethanes in Drinking Water Treatment" Ozone Science & Engineering 2(1):75-99.

Rice, R.G., C.M. Robson, G.W. Miller, and A.G. Hill, 1981, "Uses of Ozone in Drinking Water Treatment", J. Am. Water Works Assoc. 73(1):44-57.

Rice, R.G., C.M. Robson, G.W. Miller, J.C. Clark, and W. Kühn, 1982, "Biological Processes in the Treatment of Municipal Water Supplies", EPA Report No. 600/S2-82-020. Natl. Tech. Info. Service, Report No. PB 82-199,704.

Schulhof, P., 1979, "French Experience in the Use of Activated Carbon for Water Treatment", presented at NATO-CCMS Conference on Practical Applications of Adsorption Techniques in Drinking Water Treatment, Reston, VA (Apr. 30-May 2). U.S. EPA, Office of Drinking Water, Washington, DC.

Schulhof, P., 1980, "Water Supply in the Paris Suburbs: Changing Treatment for Changing Demands", J. Am. Water Works Assoc. 72(8):428.

Simcoe, W.D., 1983, "Potsdam, New York Water Treatment Plant: A Case History", Ozone Science & Engineering 5(1):51-60.

Singer, P.C., and W.B. Zilli, 1975, "Ozonation of Ammonia in Wastewater", Water Research 9:127-134.

Sontheimer, H., E. Heilker, M. Jekel, H. Nolte, and F.H. Vollmer, 1978, J. Am. Water Works Assoc. 70(7):393-396.

Stolarik, G., 1983, "Ozonation -- Direct Filtration of Los Angeles Drinking Water", presented at Intl. Ozone Assoc. 6th Ozone World Congress, Washington, DC, May.

Wagner, R., and R.A. Elefritz, 1983, "Ozonation for Effective THM Control", Public Works 114(4):46-48.

CHAPTER 11

THE OCCURRENCE OF CONTAMINATION IN DRINKING WATER FROM GROUNDWATER SOURCES

Hugh F. Hanson, P.E.

Chief, Science and Technology Branch
Office of Drinking Water
U.S. Environmental Protection Agency
Washington, DC 20460

It has been estimated that between 30 and 60 quadrillion gallons of groundwater occur within 1/2 mile of the surface in the United States -- a volume of water greater than four times that contained in the Great Lakes. Approximately 25% of the freshwater used in the United States is from groundwater sources. Although the predominant uses of groundwater are for irrigation and industrial applications, about 20% of groundwater is used for drinking water, including both public water supplies and individual, usually rural, supplies. It is estimated that about 50% of all U.S. residents obtain drinking water from groundwater. Historically, groundwater has been viewed as a relatively pristine resource, and generally has been used for drinking water without major treatment other than disinfection. In recent years, however, it has been recognized that groundwater is vulnerable to contamination, particularly with organic chemicals resulting from a range of man's industrial and agricultural activities.

The purpose of this paper is to summarize the information currently available to EPA's Office of Drinking Water on the occurrence of contaminants in public water supplies using groundwater sources.

Public water supplies, as defined in the Safe Drinking Water Act, are those serving 25 or more people. It is currently estimated that there are in excess of 45,000 public groundwater supplies serving more than 100 million people in the United States. It should be noted that there are also more than 11 million private wells providing drinking water to an additional 25-30 million residents, mainly in rural areas. The distribution of groundwater systems as related to population groups is given in Table I.

TABLE I. GROUNDWATER SYSTEMS IN RELATION TO POPULATION GROUPS

POPULATION SERVED		NO. SYSTEMS
25-100		19,200
100-150		13,400
500-1,000		4,200
1,000-10,000		7,300
10,000-100,000		1,200
100,000-1 Million		66
Over 1 Million		1
	TOTAL	45,000

The Safe Drinking Water Act gives EPA the authority to establish national standards in the form of maximum contaminant levels, or MCLs, that specify the maximum permissible level of a contaminant in water which is delivered to any user of a public water system. In December 1975, EPA promulgated the National Interim Primary Drinking Water Regulations which established MCLs and monitoring and reporting requirements for 10 inorganic materials, six organic pesticides, radionuclides, microbiological contaminants, and turbidity. In November 1979, these regulations were amended to include an MCL, monitoring and reporting requirements for total trihalomethanes (Table II).

Violations of the MCLs are reported by the states to EPA. Those data are maintained by EPA on the Federal Reporting Data System (FRDS).

With the exception of microbiological contamination, the reported violations of MCLs in groundwater supplies are relatively few. Approximately 4,500 groundwater supplies have been reported to exceed the microbiological MCL. Most of the violations are by small supplies.

It is estimated that there are between 1,500 and 3,000 groundwater supplies in violation of the MCLs for the inorganics including the metals. Most of these -- 1,000-2,000 -- are violations of the fluoride standards and about 500-600 are violations of the nitrate standard. Among the metals, violations of selenium and arsenic MCLs appear most frequently. There are scattered violation reports of the radionuclide standards, and very few (i.e., less than 10) violations of the pesticide MCLs.

The major concern with groundwater sources of drinking water at present is the occurrence of volatile organic chemicals. In March 1982, EPA issued an Advanced Notice of Proposed Rulemaking identifying 14 VOCs most commonly detected in groundwater which EPA has been considering for regulation under Phase I of the Revised National Primary Drinking Water Regulations (Table III). Some additional

TABLE II. CONTAMINANTS CURRENTLY REGULATED UNDER THE INTERIM NATIONAL PRIMARY DRINKING WATER REGULATIONS

	MCL
Inorganics	
Arsenic	0.05 mg/L
Barium	1.0 mg/L
Cadmium	0.010 mg/L
Chromium	0.05 mg/L
Lead	0.05 mg/L
Mercury	0.002 mg/L
Nitrate (as N)	10 mg/L
Selenium	0.01 mg/L
Silver	0.05 mg/L
Fluoride	1.4-2.4 mg/L (ambient temperature dependent)
Organic Pesticides	
Endrin	0.0002 mg/L
Lindane	0.004 mg/L
Methoxychlor	0.1 mg/L
Toxaphene	0.005 mg/L
2,4-D	0.1 mg/L
2,4,5,-TP (Silvex)	0.01 mg/L
Radionuclides	
Radium-226 plus Radium-228	5 pCi/L
Gross alpha particle activity	15 pCi/L
Beta particle and photon radioactivity	4 mrem/year
Turbidity	1 Tu (up to 5 Tu)
Total Trihalomethanes	0.01 mg/L
Coliform Bacteria	Varies as function of system size, sampling frequency, and analytical methods used

TABLE III. VOCS EXAMINED IN THE SIX NATIONAL SURVEYS

	NORS	NOMS	NSP	CWSS	GWSS	RWS
Approximate Number of Groundwater Systems Sampled	16	18	12	330	945	644
Trichloroethylene		•	•	•	•	•
Tetrachloroethylene		•	•	•	•	•
Carbon Tetrachloride	•	•	•	•	•	•
1,1,1-Trichloroethane		•	•	•	•	•
1,2-Dichloroethane	•	•	•	•	•	•
Vinyl Chloride		•	•			
Dichloromethane		•	•		•	
Benzene		•	•	•	•	
Chlorobenzene			•	•	•	
Dichlorobenzene(s)		•	•		•	
Trichlorobenzene(s)		•	•			
1,1-Dichloroethylene			•		•	
cis - or _trans_-1,2-Dichloroethylene			•	•	•	•

organic contaminants (e.g., pesticides like DBCP and EDB) are being considered for Phase II of the revised regulations. Because there are no MCLs or monitoring requirements currently for VOCs, a comprehensive data base on the occurrence of VOCs in public water supplies does not exist. Estimates of the national occurrence of VOC's have been developed from monitoring data obtained through extrapolation from six national surveys conducted over the last several years:

- National Organics Reconnaissance Survey (NORS)
- National Organics Monitoring Survey (NOMS)
- National Screening Program for Organics in Drinking Water (NSP).
- Community Water Supply Survey (CWSS)
- Rural Water Survey (RWS)
- Groundwater Supply Survey (GWSS)

The number of groundwater supplies sampled and the particular VOC examined in each survey are also shown in Table III.

Table IV shows the estimated percentages of groundwater supplies nationally that are contaminated with various VOCs as extrapolated from the data available from the six surveys. In general, the frequency of occurrence is much greater among larger supplies, although in absolute numbers, more small supplies are expected to be contaminated. Table V summarizes data obtained on VOC occurrences and their maximum concentrations found.

TABLE IV. ESTIMATED NATIONAL FREQUENCY OF OCCURENCE OF VOCS IN GROUNDWATER SUPPLIES

VOC	0.5 microg/L	5 microg/L	20 microg/L	50 microg/L	100 microg/L
Trichloroethylene	3.4%	0.9%	0.4%	0.3%	0.1%
1,1,1-Trichloroethane	2.9%	0.8%	0.2%	0.01%	0.01%
Tetrachloroethylene	3.2%	0.7%	0.1%	<0.01%	
Benzene	1.3%	0.3%	0.04%	<0.01%	
Vinyl Chloride	0.06%	0.04%	<0.01%	<0.01%	
<u>cis/trans</u>-Dichloroethylene	1.1%	0.4%	0.02%		
1,1-Dichloroethylene	1.8%	0.2%			
Carbon Tetrachloride	0.7%	0.2%			
p-Dichlorobenzene	1.0%				
1,2-Dichloroethane	0.3%				
Chlorobenzene	0.2%				
o-Dichlorobenzene	0.2%				

TABLE V. SUMMARY OF VOC FINDINGS IN THE GROUNDWATER SURVEY

CHEMICAL	#STATES TESTED	#WELLS* TESTED	%POSITIVE**	MAXIMUM MICROG/L
Trichloroethylene	8	2894	28	35,000
Carbon Tetrachloride	4	1659	10	379
Tetrachloroethylene	5	1652	14	50
1,2-Dichloroethane	2	1212	7	400
1,1,1-Trichloroethane	3	1611	23	401,300
1,1-Dichloroethane	9	785	18	11,300
Dichloroethylenes (3)	8	781	23	860
Methylene Chloride	10	1183	2	3,600
Vinyl Chloride	9	1033	7	380

* Ratio of community wells to private wells is not known
** Not a statistical value

Although the data suggest that fewer than 5% of the nation's groundwater supplies have detectable levels of any one specific VOC, an analysis of the data from about 500 randomly selected samples of the Groundwater Supply Survey (GWSS) (Figure 1) indicated that 21.2% of the nation's supplies may have at least one VOC present at detectable levels, and that 5.6% have three or more present.

RESULTS OF ANALYSES:

Random - 21% of systems were positive

Non-Random - 27% of systems were positive

SYSTEMS SERVING 10,000 PEOPLE OR MORE FOUND TO HAVE HIGHER FREQUENCY OF CONTAMINATION:

10,000 - 16.5%

10,000 - 28.5%

VOC CONTAMINATION CANNOT BE PREDICTED WITH MUCH ACCURACY BASED UPON PROXIMITY TO INDUSTRIES, LANDFILLS, OR OTHER POTENTIAL SOURCES.

Figure 1. Groundwater Supply Survey results.

Conversely, a non-random selection of sample sites taken in the GWSS indicated a higher frequency of occurrence (27%) and much higher concentrations.

Groundwater contamination with VOCs, when found, is detected most commonly at less than 10 mg/L with smaller percentages in the 10-100 microg/L and 100-1000 microg/L range, respectively. By contrast, VOC contamination of surface water, when found, is detected most commonly at less than 5 microg/L.

It is very clear that the distribution of organics in drinking water is directly related to use patterns of the specific chemicals. This is even more dramatic in the case of heavy molecular weight chemicals, such as pesticides and herbicides. Therefore, EPA will be analyzing all of these occurrence predictors in order to develop monitoring strategies that can effectively identify and quantify groundwater contamination, and will propose those strategies as part of the pending VOC proposal and the anticipated Advanced Notice of Proposed Rulemaking for the revision of the National Interim Primary Drinking Water standards. The Agency believes that the problems in groundwater can be identified and mitigated through effective and highly qualified monitoring. The present efforts of the Office of Drinking Water are concentrating on the development of revised regulations which will specify the appropriate monitoring regimens.

CHAPTER 12

IMPROVED MONITORING TECHNIQUES TO ASSESS GROUNDWATER QUALITY NEAR SOURCES OF CONTAMINATION

Glenn E. Schweitzer

Director
Environmental Monitoring Systems Laboratory
U.S. Environmental Protection Agency
Las Vegas, Nevada 89114

APPROACHES TO GROUNDWATER QUALITY ASSESSMENT

Concurrent with the increased interest in protecting the quality of groundwater has been an expanded effort to improve our capabilities for identifying and quantifying chemical contaminants in groundwater. Major objectives have included reductions in the cost, time and uncertainties involved in groundwater monitoring. The results to date of recent research and development activities directed to these objectives have been encouraging, and such activities undoubtedly will continue to increase in size and scope during the next several years.

Much of our information about groundwater quality has come from monitoring surveys of contaminants entering drinking water systems that use groundwater and surveys of contaminants in tap water that is drawn from individual wells. Unfortunately, however, this approach of sampling at the point of use may uncover problems only after an aquifer has been irreversibly contaminated.

A second approach to assessing groundwater quality is to examine all available monitoring data about an aquifer, identify contaminant sources of the aquifer and hydrogeological conditions of the area, and then attempt to characterize the quality of the aquifer. In some areas of the country monitoring wells have been in place for many years and can provide data for such assessments, although coverage has been very spotty both in terms of geographic distribution and chemicals of interest. Recently, several states have initiated more systematic monitoring programs in some areas. They have installed a number of wells and are conducting analyses for a range of pesticides and other organic compounds to provide an improved data base on groundwater contamination.

A third approach, which is the focus of this presentation, is monitoring near the sources of groundwater contamination. The objectives are to uncover potential pollution problems as early as possible and, by relating contamination levels to sources, to provide a basis for preventive or corrective actions. In this regard the Resource Conservation and Recovery Act (RCRA) and the Superfund legislation clearly are a major impetus for increased groundwater protection efforts, particularly as subsurface contaminants are discovered near many abandoned disposal sites. Also, the increasing frequency of persistent agricultural chemicals showing up in groundwater in agricultural areas and the widespread concerns over petroleum products leaking from underground storage tanks at refineries, airports, and active and abandoned filling stations have underscored the importance of monitoring near the sources of pollution.

Reducing the costs of laboratory analyses is a central concern in every approach to groundwater monitoring. While costs of routine analyses for large numbers of chemicals have dropped dramatically during the past several years, the costs still are substantial. A variety of field screening procedures now is being used, and additional field and laboratory screening techniques are being developed in efforts to reduce the need for detailed analyses of all samples that are collected.

USING AERIAL PHOTOGRAPHY TO IDENTIFY GEOGRAPHICAL AREAS OF CONCERN

During the past several years, surveys for active and abandoned hazardous waste sites have been carried out in almost every part of the country. Most of the larger sites probably have now been located by state and local authorities. However, many of these sites have not yet been characterized in even a preliminary manner, and there may be a few sites of major significance which remain undiscovered beneath overgrown vegetation or intermingled within man-made urban or industrial developments.

Aerial photography has been a very valuable tool in identifying, characterizing, and cataloging sites. In many cases photos taken 10 to 40 years ago have provided the initial clues as to the presence of abandoned sites. Photography can play an important role in assessing the likelihood that chemicals are migrating away from a site. Photos may provide direct evidence of migration pathways by identifying vegetation damage or runoff patterns. They may confirm that containment has not been breached or that cleanup efforts have been successful. Probably most often they identify locations that warrant detailed on-site investigations.

The coupling of aerial photos showing potential sources of groundwater contamination and possible migration pathways with information on the surficial geology and the location and characteristics of nearby aquifers would further clarify the potential for groundwater contamination. Unfortunately, adequate information on the microgeology of individual sites seldom is available, and subsurface profiles must be developed during the course of environmental investigations. However, once such profiles are in hand, preparation of georeferenced data sets in digital form linking the profiles and aerial imagery would offer a number of possibilities for integrating a variety of indicators of contamination problems and for presenting such integrated data in a comprehensible manner. Similar techniques have been developed for integrating a variety of parameters affecting large geographical areas, but have yet to be adapted to smaller areas associated with specific point sources of groundwater contamination represented by waste sites.

While characterizing sites used exclusively for hazardous waste disposal presents difficult challenges, more complex problems are encountered in the identification and characterization of chemical disposal activities that have not been confined to such sites. Specifically, co-mingling of chemical wastes with municipal refuse at city landfills has been common in many localities. Abandoned underground storage tanks in industrial areas and at thousands of service stations throughout the country now are emerging as major sources of underground contamination. Finally, the current efforts in a number of states to ban or severely limit the land disposal of chemical wastes may be resulting in a revival of midnight dumping practices. The painstaking process of ferreting out these types of problems is underway. While aerial photos may be of only limited use in these efforts, frequently they can provide at least a point of departure for other investigative techniques.

A specialized use of aerial photography, and particularly historical photography acquired in the 1930s and 1940s, has been the identification of abandoned oil wells that could provide man-made pollution conduits into aquifers. The presence in the United States of more than 1.8 million active and abandoned wells underscores the importance of this concern. Initial efforts have demonstrated a very high discovery rate of abandoned wells using photos and the feasibility of pinpointing the well locations within six feet for ground investigations. Even in cases where county records indicate the presence of such abandoned wells, the records and available maps frequently are not sufficiently accurate to allow such pinpointing, or indeed even discovery in fields that are now overgrown.

DELINEATING CONTAMINANT PLUMES IN GROUNDWATER

Progress is being made in adapting several surface geophysical techniques developed primarily for oil and mineral exploration to the identification of inorganic contaminant plumes in groundwater. Field investigations of a contaminated aquifer using electromagnetic induction and resistivity techniques, in particular, show good correlation between the results obtained with these techniques and analyses of groundwater samples from monitoring wells. The next steps are to improve the calibration of the geophysical methods and to document more fully their capabilities and limitations considering different depths, geological conditions, and contaminant targets.

Downhole sensing techniques also should be helpful in discovering and delineating inorganic contaminant plumes, particularly with regard to vertical profiles. The depths of principal interest are 10 to 100 feet, with a maximum depth of 300 feet. Resistivity and gamma logging are among the most promising techniques being investigated. Existing monitoring wells for implantation of the probes are used whenever possible, although in some cases drilling small diameter boreholes may be appropriate. Downhole sensing also is important in characterizing the subsurface stratigraphy and the speed and direction of groundwater flow which, in turn, provide essential insights as to plume migration.

The problem of remotely sensing organic plumes is more difficult. One approach is to assume that some inorganic contaminants tend to move together with the organics of concern. Geophysical techniques can locate the inorganics and thus the general area of the plume. Then groundwater sampling, or perhaps soil gas sampling near the surface, using Draeger tubes or surface or subsurface flux traps, may be able to delineate the organic contamination. There have been successful efforts to detect hydrocarbons directly using electromagnetic induction. However, the electrical conductivity phenomena are not well understood, and the technique is applicable only in certain types of subsurface environments.

Usually, groundwater samples will provide the authoritative information needed to characterize aquifer contamination with remote monitoring techniques used to guide the location of monitoring wells and thereby reduce costs of the sampling program. Frequently, wells already will be in place and the geophysical data will help determine the appropriateness of relying on these wells and whether additional locations are in order.

STATISTICAL BASIS FOR SAMPLING

Underlying the approach to groundwater sampling should be a statistical basis for acquiring representative data.

The statistical approach, in turn, depends on the specific objectives of the monitoring program, e.g., characterize the extent of the plume in the aquifer, detect changes above background which indicate the presence of a plume attributable to a man-made source, or compare contamination levels in a contaminated area with contamination levels in a control area. Implementing sampling schemes based on statistical considerations, which usually call for many sampling points, may be costly, but the objectivity provided by statistics can provide very useful guidance for acquiring and interpreting monitoring data.

In operational settings, cost constraints may limit the number of wells and sample analyses to such an extent that the application of statistics is not practical. However, as the costs of remedial actions increase, and with increased reliance on monitoring data as the principal basis for such actions, larger expenditures on monitoring increasingly are warranted.

Geostatistics have found increasing receptivity in efforts to characterize the areal extent of surface soil contamination near point sources. Unlike statistics based on random variables, this technique recognizes that contaminant levels in samples taken close together are more likely to be similar than in samples taken farther apart. Using data from preliminary sampling, geostatistics have provided an objective basis for determining appropriate sample spacing. This technique also provides a basis for interpolating contaminant levels between levels measured at sampling points. Such interpolations facilitate the preparation of maps which present isopleths which connect points with the same levels of contamination. These maps then can be used in determining the localized areas of high contamination that may warrant cleanup activities.

Evaluation of the appropriate spatial location of groundwater monitoring wells is often a complicated problem not entirely amenable to solution by classical statistics. Groundwater and contaminant movement is completely determined by physical parameters such as permeability, porosity, piezometric head gradients, fault location and orientation, and other factors. The effect of diffusion is very small. Therefore, contaminant migration is not a random process; in statistical terms, it is characterized by high spatial correlation.

Application of geostatistics to this three-dimensional problem of groundwater contamination now is being explored. Given the costs of installing monitoring wells, the amount of data required for successfully using this technique is of particular concern. Nevertheless, if authoritative three-dimensional isopleths of aquifer contamination could be obtained, the costs of acquiring the necessary input data well may be warranted.

In comparing the means of contamination levels between samples taken up-gradient and down-gradient of a known pollution source, classical statistics may provide a basis for determining whether contamination can be attributed to the source. Several standard statistical tests are available for comparing mean contaminant values from the samples that are taken. The statistical tests to be used, in turn, should influence the number of samples that are collected -- either in terms of location of sampling points and/or sampling frequency.

Similarly, statistics are important in determining whether levels of contamination that are reported exceed prescribed standards. Of particular concern are those cases in which there are technical uncertainties associated with the chemical analyses, and the reported levels are close to the standards. Statistics can provide a basis for established confidence limits that levels reported above or below the standards are in fact above or below.

IMPROVED SAMPLING AND ANALYSIS TECHNIQUES

As groundwater monitoring activities expand to include more chemicals at lower concentration levels, the possible inadvertent introduction of contaminants into monitoring wells during drilling or sampling is of growing concern. The presence or lack of hydraulic connections between shallow and deeper aquifers may be a key question as to containment, and care is needed to insure that monitoring wells do not inadvertently provide such connections. Also, drilling techniques must allow for appropriate grouting as the drill bit passes from one aquifer to another if accurate vertical profiles of contaminants are to be obtained.

In addition to cross contamination between aquifers, the contamination potential from the materials used during drilling, casing, and sampling is of concern. Specifically, the materials that can be used in casings (e.g., polyvinyl chloride, polyethylene, Teflon, stainless steel) and in submersible pumps (e.g., Teflon, glass, stainless steel) are receiving greater scrutiny. Greater care is needed in the selection of cements, glues, and cleaning agents.

With regard to methods for analyzing the large number of groundwater samples that are being collected, costs are a major concern. At present the analytical protocols being used for the Superfund program call for quantitation of a large number of organic and inorganic chemicals in the 10-100 ppb range at a cost at commercial laboratories of about $700-$800 per sample. While this cost is greatly reduced from costs several years ago, the price still is too high for routine analyses of samples that are not expected to be contaminated.

Field screening techniques now are being used for detecting volatile organic compounds and halogenated pesticides which are found to be associated with many hazardous waste sites. Such field techniques provide data with minimal delays, as well as reducing the analytical costs by identifying those samples that warrant detailed laboratory analysis. One of the most widely used approaches is Organic Volatile Analysis, which allows headspace vapor samples to be injected into a flame ionization detector. If volatiles are detected, injections then are made onto a portable gas chromatographic column. Standards are used in the field to allow quantitation in the 10-100 ppb range. Electron capture detectors, in combination with gas chromatographs, provide a basis for measuring pesticides and polychorinated biphenyls in the one ppm range.

A number of mobile laboratories also are being used for more detailed analyses in the field. Most recently, tandem mass spectrometry units are being used in addition to the more standard laboratory techniques that have been taken to the field.

Finally, the use of fiber optics technology in combination with laser fluorescence spectroscopy as a means for detecting groundwater contaminants in the field has attracted considerable attention in recent months. The concept calls for transmitting a laser signal through a cable inserted into a small diameter borehole to the appropriate depth and then correlating the return signal with calibrated fluorescence characteristics of the contaminants. Current work is being directed to development of an optical-chemical transducer that will permit sensing of inorganic chlorides. The next step will be to develop a transducer for organic chlorides. Then it should be possible to prepare fluorescence response curves that can be used for measuring concentrations of selected compounds, which could serve as indicator chemicals.

WHAT CHEMICALS SHOULD BE MONITORED AND AT WHAT LEVELS?

As a minimum, those chemicals that are regulated under the Safe Drinking Water Act usually are of interest in investigations of possible groundwater contamination. At the other extreme, federal legislation has been proposed which would require monitoring for "all" chemicals that might be present. In between, some states have their own chemical "hit lists." For example, one state requires monitoring for 80 parameters in some programs. An approach that seems to be gaining increasing attention is to monitor for the priority pollutants, since the procedures for analyzing for these 129 chemicals as a package have been well developed.

The groundwater monitoring requirements under RCRA are developing rapidly. The current approach calls for two types of groundwater monitoring: (1) detection monitoring to determine whether contaminants reach the groundwater, and if they do, (2) compliance monitoring to determine whether proscribed levels are being exceeded.

For detection monitoring, a set of indicator parameters has been proposed for monitoring at all sites (Total Organic Halogen, Total Organic Carbon, pH, temperature, specific conductance), with EPA specifying other hazardous constituents of interest on a site-by-site basis. Among the factors to be considered in specifying the additional constituents are the mobility, stability, and persistence of the chemicals deposited at the site and of their degradation products; the detectability of the chemicals of concern; and the background presence of the chemicals in the groundwater due to natural or man-made sources of contamination other than the water site of interest.

Monitoring requirements at Superfund sites are less well defined, given the wide variety of situations that are encountered. Such requirements are incorporated increasingly into consent decrees and undoubtedly will become a major concern of the Post-closure Liability Trust Fund which will administer many monitoring programs following cleanup. One current approach is to specify the chemicals of concern based on the results of preliminary site assessments, and particularly analyses of leachates discovered to have migrated away from the site. Another approach is to monitor for a standard hit list of chemicals (currently about 150) and, in addition, to identify those other organic chemicals that show significant peaks on a gas chromatograph/mass spectrometry system in addition to the peaks indicating the presence of hit list compounds.

Determination of the concentration levels of contaminants in groundwater that should be of concern is central to the development of effective monitoring programs. One aspect of such determinations is the sensitivity of available analytical methods, although health concerns presumbly present the overriding considerations.

At present, several approaches are being used to set contaminant levels of concern. Obviously, drinking water standards, health advisory levels, and related numerical limitations developed by the Federal or State Governments are the core action levels for aquifers which serve as drinking water sources. For other contaminants in such aquifers, or for all contaminants in aquifers which have no potential as sources of drinking water, an approach is to set the lowest levels that are technically feasible to attain. This approach introduces consideration of the costs of cleaning up the aquifer, as well as the costs of altering activities contributing to aquifer contamination. A third approach is to declare that background levels in the aquifer are the appropriate action levels and that contaminant sources which increase these levels are prohibited or must be cleaned up. Of course, the problems in defining background are formidable.

SUMMARY

Our approach to groundwater protection is critically dependent on our ability to determine the present and potential extent and seriousness of contamination patterns.

The aquifers are so extensive and the costs of assessing subsurface pollution so high that limited resources must be carefully husbanded. We cannot afford to make measurements that lack credibility or meaning, nor can we afford not to make important measurements.

CHAPTER 13

NBS ENVIRONMENTAL STANDARD REFERENCE MATERIALS FOR USE IN VALIDATING WATER ANALYSIS

R. Alvarez

Office of Standard Reference Materials
United States Department of Commerce
National Bureau of Stanards (NBS)
Washington, DC 20234

In recent years, the analysis of drinking water has emphasized the identification and determination of specific chemical pollutants which represent public health hazards. A number of these priority pollutants are present at microg/L levels. For use in calibrating instrumentation and validating experimental data and methods, NBS issues Standard Reference Materials (SRMs) certified for concentrations of important inorganic and organic priority pollutants. Three water SRMs are available, which are certified for inorganic contaminants. They are: SRMs 1641b and 1642b, which are certified for mercury at the microg/mL and ng/mL levels, respectively; and SRM 1643a, Trace Elements in Water, which is certified for 17 trace elements.

Three SRMs also have been issued for organic pollutants. They are: SRM 1644, Generator Columns for Polynuclear Aromatic Hydrocarbons, which provides a means of generating known, accurate aqueous concentrations of anthracene, benz(a)anthracene, and benzo(a)pyrene; SRM 1647, Priority Pollutant Polynuclear Aromatic Hydrocarbons (in Acetonitrile); and most recently, SRM 1639, Halocarbons (in Methanol) for Water Analysis, which is certified for the concentrations of four trihalomethanes and two other halocarbons.

The Certificate of Analysis for SRM 1639 lists certified concentrations of chloroform, chlorodibromomethane, bromodichloromethane, bromoform, trichloroethylene, and tetrachloroethylene. The certified concentrations of these volatile compounds are based on gas chromatography and on the mass of the halocarbon, corrected for compound purity, added to the methanol. SRMs 1639 and 1647 are intended primarily for adding accurate amounts of the certified compounds to water and determining response factors.

Copies of SRM 1639 [Halocarbons (in Methanol) for Water Analysis] and SRM 1643a [Trace Elements in Water], are attached to this presentation.

Other SRMs are being developed. For example, an SRM will be certified for the concentrations of approximately ten phenols in methanol. Another SRM is being developed for use with combined gas chromatography/mass spectrometry. It will consist of two solutions: one certified for the concentrations of approximately ten priority pollutants; the other for the concentrations of the same compounds which have been labeled isotopically. The compounds will represent acid extractable, base-neutral extractable, and volatile classes.

NBS is authorized by Federal legislation to issue SRMs which have important applications in quality assurance. We welcome your suggestions for additional SRMs that may be needed to improve measurement reliability.

National Bureau of Standards

Certificate of Analysis

Standard Reference Material 1639

Halocarbons (in methanol) for Water Analysis

This Standard Reference Material is intended primarily for calibrating chromatographic instrumentation used in the determination of halocarbons. It is also useful in recovery studies for adding accurate amounts of the certified compounds to a sample. Because of its miscibility with water, it is particularly useful in analyzing water samples for these compounds.

Certified Concentrations of the Halocarbons: The certified concentrations and estimated uncertainties of seven halocarbons in methanol are shown in Table 1. Because the density of methanol changes with temperature, these concentrations are certified for the temperature range of 20 to 26 °C.

They are based on the analytical results by gas chromatography (GC) and on the concentrations calculated from the mass of the halocarbon added, corrected for compound purity, to methanol. In addition to GC, high performance liquid chromatography (HPLC) was used to confirm the homogeneity of the entire lot. Table 2 shows the calculated and GC-determined concentrations.

NOTICE AND WARNINGS TO USERS

Expiration of Certification: This certification is valid, within the limits certified, for one year from the date of purchase. In the event that the certification should become invalid before then, purchasers will be notified by NBS.

Storage: Sealed ampoules, as received, should be stored in the dark at temperatures between 10-30 °C.

Use: Samples of the SRM for analysis should be withdrawn from ampoules equilibrated at 23 ± 3 °C. Samples should be withdrawn immediately after opening and used without delay for any certified value in Table 1 to be valid within the stated uncertainty. Certified values are not applicable to ampoules stored after opening, even if resealed. A suggested procedure for preparing and equilibrating water calibration standards is described in a separate section.

Analytical determinations were performed at the Center for Analytical Chemistry, Organic Analytical Research Division, by S.N. Chesler, R.G. Christensen, and F.R. Guenther.

The statistical design and analysis of the experimental work was provided by K. Kafadar and K.R. Eberhardt of the Statistical Engineering Division.

The coordination of the technical measurements leading to certification was performed under the direction of F.R. Guenther, S.N. Chesler, and H.S. Hertz.

The technical and support aspects involved in preparation, certification, and issuance of this Standard Reference Material were coodinated through the Office of Standard Reference Materials by R. Alvarez.

Washington, DC 20234
April 20, 1983

George A. Uriano, Chief
Office of Standard Reference Materials

(over)

Table 1. Certified Concentration of Halocarbons in SRM 1639 at 23 ± 3 °C

Compound	Concentration, ng/μL*
Chloroform	6235 ± 340
Chlorodibromomethane	124.6 ± 1.1
Bromodichloromethane	389.9 ± 7.1
Bromoform	86.5 ± 1.4
Carbon tetrachloride	157.0 ± 4.4
Trichloroethylene	85.8 ± 2.6
Tetrachloroethylene	40.6 ± 0.9

*Estimated uncertainty is given as 95% confidence limits obtained from the GC measurements.

Table 2. Summary of Results

Compound	Concentration, ng/μL	
	Calculated[a]	GC[b]
Chloroform	6311 ± 34	6235 ± 338
Chlorodibromomethane	124.9 ± 0.1	124.6 ± 1.1
Bromodichloromethane	392.0 ± 0.3	389.9 ± 7.1
Bromoform	87.7 ± 0.1	86.5 ± 1.4
Carbon tetrachloride	160.1 ± 0.1	157.0 ± 4.4
Trichloroethylene	87.9 ± 0.1	85.8 ± 2.6
Tetrachloroethylene	40.95 ± 0.03	40.6 ± 0.9

[a]The calculated concentration is based on the total mass of the halocarbon added to the methanol.
[b]Estimated uncertainty is given as 95% confidence limits.

Suggested Procedure for Preparing and Equilibrating Water Calibration Standards

The following procedure provides aqueous halocarbon concentrations in the range normally found in drinking water supplies.

1. Allow ampoule to equilibrate at a temperature of 23 ± 3 °C and shake ampoule for one minute.
2. Open ampoule and complete sampling of contents within five minutes.
3. Insert one-μL disposable micropipet (such as "Microcap"* available from laboratory supply houses) into ampoule and fill micropipet from top of SRM solution.
4. Inject the one-μL volume into a stoppered 100-mL volumetric flask containing 100-mL pure water and rinse the micropipet twice with this solution.
5. Shake the flask for at least one minute and analyze entire sample. However, because the one-minute shaking time may not have achieved a uniform concentration of halocarbons in the water, the analysis of aliquots may require a longer shaking time.

Other concentrations can be obtained by the use of other calibrated micropipets and/or volumetric flasks of suitable volume.

Preparation and Analysis of SRM 1639

The methanol solution of the seven halocarbons was prepared at NBS. It was chilled and ampouled into 2-mL amber glass ampoules. The ampoules were purged with argon immediately before adding the solution. Samples representing early, middle, and final stages of ampouling were analyzed by GC. HPLC was also used to confirm homogeneity of the material for the SRM. No significant differences in concentrations of the seven compounds were found.

GC analyses were done on a 30 m x 0.25 mm inside diameter fused silica capillary column coated with a one-μm thick immobilized SE-52* phase. Bromotrichloromethane was used as an internal standard. A Hall electrolytic conductivity detector was used in the halogen mode for the determination of all constituents except bromoform and chlorodibromomethane. These highly brominated compounds were determined using electron capture detection.

*Note: To describe the procedure adequately, identification of a commercial product by the manufacturer's name was necessary. This identification does not imply endorsement by NBS nor that the particular product is the best available.

U. S. Department of Commerce
Malcolm Baldrige
Secretary

National Bureau of Standards
Ernest Ambler, Director

National Bureau of Standards

Certificate

Standard Reference Material 1643b

Trace Elements in Water

This Standard Reference Material (SRM) is intended primarily for use in evaluating the accuracy of trace element determinations in filtered and acidified fresh water and for calibrating instrumentation used in these determinations. SRM 1643b consists of approximately 950 mL of water in a polyethylene bottle, which is sealed in an aluminized bag to maintain stability. SRM 1643b simulates the elemental composition of fresh water. Nitric acid is present at a concentration of 0.5 mole per liter to stabilize the trace elements.

Concentrations of Constituent Elements: The concentrations of the trace elements that were determined are shown in Table 1. The certified values are based on results obtained either by reference methods of known accuracy or by two or more independent, reliable analytical methods. Noncertified values, which are given for information only, appear in parentheses.

Notice and Warnings to Users:

Expiration of Certification: This certification is invalid two years after the shipping date.

Precautions: The bottle should be shaken before use because of possible water vapor condensation. To prevent possible contamination of the SRM, do not insert pipets into the bottle. After use, the bottle should be capped tightly and placed inside the aluminized bag, which should be folded and sealed with sealing tape. This safeguard will protect the SRM from possible environmental contamination and long-term loss of water.

Elemental determinations of ng/g levels are limited by contamination. Apparatus should be scrupulously cleaned and only the purest grade reagents employed. Sampling and manipulations, such as evaporations, should be done in a clean environment, for example, a Class 100 clean hood.

The overall direction and coordination of the technical measurements leading to this certification were performed under the direction of E. Garner, Chief of the Inorganic Analytical Research Division.

The technical and support aspects involved in the preparation, certification, and issuance of this Standard Reference Material were coordinated through the Office of Standard Reference Materials by R. Alvarez.

Washington, DC 20234
May 18, 1984

Stanley D. Rasberry, Chief
Office of Standard Reference Materials

(over)

(Table 1)

Concentrations of Constituent Elements

Element	Concentration,* ng/g	Element	Concentration,* ng/g
Arsenic[1,5]	(49)**	Lead[3,4b]	23.7 ± 0.7
Barium[2a,2b,5]	44 ± 2	Manganese[1,2a,3]	28 ± 2
Beryllium[1,2a]	19 ± 2	Molybdenum[2a,5]	85 ± 3
Bismuth[1]	(11)	Nickel[2a,3]	49 ± 3
Boron[2a]	(94)	Selenium[1,5]	9.7 ± 0.5
Cadmium[2b,3,5]	20 ± 1	Silver[1,5]	9.8 ± 0.8
Chromium[4b]	18.6 ± 0.4	Strontium[2a,5]	227 ± 6
Cobalt[1,5]	26 ± 1	Thallium[4b]	8.0 ± 0.2
Copper[3,4b]	21.9 ± 0.4	Vanadium[4b]	45.2 ± 0.4
Iron[2a,4a,5]	99 ± 8	Zinc[2a,5]	66 ± 2

* The estimated uncertainty is based on judgment and represents an evaluation of the combined effects of method imprecision and possible systematic errors among methods. To convert to nanograms per milliliter, multiply by the density of the SRM. The density at 23 °C is 1.017 grams per milliliter.

** Values in parentheses are not certified.

1. Atomic absorption spectrometry, electrothermal
2. Atomic emission spectrometry,
   a. dc plasma
   b. flame
3. Laser enhanced ionization flame spectrometry
4. Isotopic dilution mass spectrometry,
   a. resonance ionization
   b. thermal ionization
5. Neutron activation, instrumental

Source and Preparation of Material: SRM 1643b was prepared at the U.S. Geological Survey, National Water Quality Laboratory, Arvada, Colorado, under the direction of V.J. Janzer of that laboratory and J.R. Moody of the NBS Center for Analytical Chemistry. Only high-purity reagents were used and the containers were acid-cleaned and sterilized before use. In the preparation, a polyethylene cylindrical tank was filled with distilled water and sufficient nitric acid to make the solution approximately 0.5 moles HNO_3 per liter. Solutions containing known amounts of calcium, sodium, magnesium, potassium, and the elements to be determined were added to the acidified water solution with constant stirring. After thoroughly mixing, the solution was filtered, sterilized, and then transferred to one-liter polyethylene bottles. The approximate concentrations, in μg/mL, of Ca, Na, Mg, and K are respectively 35, 8, 15, and 3.

Analysts:

Center for Analytical Chemistry, National Bureau of Standards

1. K. A. Brletic
2. T. A. Butler
3. E. C. Deal
4. M. S. Epstein
5. J. D. Fassett
6. K. Fitzpatrick
7. H. M. Kingston
8. R. M. Lindstrom
9. L. A. Machlan
10. J. R. Moody
11. L. J. Powell
12. T. C. Rains
13. T. A. Rush
14. S. F. Stone
15. G. C. Turk
16. R. L. Watters, Jr.
17. R. Zeisler

CHAPTER 14

REGULATION OF CONTAMINANTS IN DRINKING WATER

Joseph A. Cotruvo, Ph.D.

Director, Criteria and Standards Division
Office of Drinking Water
Environmental Protection Agency
Washington, DC 20460

It is a pleasure for me to be here and to be talking with you about drinking water quality issues. These issues are very complicated of course, and it is very difficult to divorce the health concern from the fact, from the hypothesis, from the potential. But nevertheless, all of us feel that it is essential that there be very high quality drinking water available to the public, that there should not be any question in people's minds as to the quality of their drinking water, and that it is our collective responsibility, as water quality professionals, to assure that is the case. We provide the treatment and the controls over contamination at the source, and we are involved in establishing quality specifications. Together, we must assure that the public does not have any significant risk from exposure to drinking water contamination, and that the public should be confident that their drinking water is safe.

There is a wide variety of substances in drinking water and in some cases, there are indeed clear imminent hazards; fortunately, those cases are relatively few. But there are also other cases where there is potential for concern or risk, and there are others where this is relatively speculative. And so rather than lumping them all together as hazards, such that every single substance once found automatically is concluded to be a hazard, I think we must make some differentiation. We have to deal with risks that we understand in a particular context, and those that we do not know very much about in another context.

In general we have to conclude that by all reasonable measures the drinking water is safe. Public water supplies are safe. Now, obviously, that does not mean that all of them are absolutely, and unequivocably free from every possible contaminant or every possible risk that could be there. We know that there are examples where there is very significant contamination, and we know that groundwater is a matter of increasing concern as more of these detections occur. That is real! We must deal with those. But that does not mean that people should approach their glasses of drinking water with fear and trepidation in the vast majority of cases. It also does not mean that we should not be concerned that drinking water potentially can become contaminated. I think that our main concern in the U.S. government, is to be sure that we do not permit contamination to occur, that we prevent it, that we reduce it whenever we can; and we do that aggressively! But the proper perspective also is important.

Safety is not a precise term. We would have to say that our safety in this room right now is not precarious in the sense of an airplance crashing into it in the next hour. That is a very low probability event. None of us are concerned about that happening, but nevertheless, it is important that the airlines take special precautions to ensure that they fly safely. Any that is what we are doing in drinking water: we are ensuring - as a goal - that drinking water is as good as it can be!

The other thing we should keep in mind is that except in relatively unusual cases, drinking water is not the principal or even a major source of exposure to most environmental contaminants, either industrial or otherwise. Air happens to be a very significant source, both from occupational exposure and from indoor contamination. Food can be a significant source of natural and synthetic toxicants. Drinking water is uniquely the most significant source of those substances that are associated intimately with drinking water production, especially disinfection and other treatment processes. So, obviously that requires special attention because we have complete control over it. We do not have complete control over the other sources of contamination, however; thus we still have to be diligent.

The nervousness that develops in this area is that as analytical science keeps improving, we keep detecting more and more things in drinking water, and everywhere else in the environment. The problem is that analytical chemistry is running about three, four or five orders of magnitude ahead of the ability of toxicologists and epidemiologists to determine the consequences of that exposure. So we have many more questions than we will ever be able to answer. I do not think, and I cannot even conceive under the most exceptional developments of science that one will ever be able to say with any degree of certainty what the consequences are of exposure to small amounts of various substances individually, let alone in combination. So we will never answer that question, ultimately, absolutely!

The traditional approach, as we are told from the water industry, has been to deal with problems in a type of best practical/available technology process, and is probably the best one to continue. In 1900, or near that time when chlorine was just beginning to be used in the disinfection of water, if the scientists had taken over and determined that before one could make that adjustment to treatment processes, we first had to identify each of the particular microorganisms that could possibly be in drinking water and then found its particular potency and frequency, and found its particular susceptibility to treatment and chlorination, we would probably still be wondering whether or not chlorine should be added to drinking water. The point is, someone made an excellent gut judgement in that time - and it saved millions of lives. There were lots of uncertainties. But it worked. It was cheap. It solved the problem conceptually, as opposed to dealing with the problem in its minute sense. And, in a sense, we need to keep thinking along those lines. The engineers need to think along those lines as they are designing water treatment systems. It is not so important necessarily to deal with specific individual chemicals, it is better to deal with classes and groups of contaminants as they design their systems and to optimize them to produce high quality water.

INTERIM DRINKING WATER REGULATIONS

In the rest of the presentation I will give you an idea of the activities going on in the EPA in implementing the Safe Drinking Water Act: how we are evaluating our current regulatory activities; what we are going to be doing over the next couple of years; and our general perception of the significance of certain kinds of contaminants that are found in drinking water.

The existing interim regulations in the United States involve approximately 40 substances. They are very comprehensive, although not totally encompassing, and include the broad spectrum of contaminants that can be in drinking water. They include the traditional biological contaminants, coliforms and turbidity; a spectrum of inorganic contaminants that are still most frequently found in drinking water from various sources; six organic pesticides; radionuclides, and finally trihalomethanes. In addition to those primary standards, there is a series of secondary standards that are not dealt with as matters of public health concern, but more as a matter of aesthetics.

REVISED REGULATIONS

Our management scheme for reviewing and revising all of the existing regulations is controlled by the fact that there is such a tremendous number of additional substances that must be considered in the revisions. The first time around, the Congress told us to concentrate on those substances that were in the U.S. Public Health Service guidelines; we did that, and extended them somewhat to the trihalomethanes and radionuclides. But since that time, analytical chemistry has been so intensively applied in drinking water systems around the U.S. that we found more substances and, so, we have to consider them for regulation. The task was divided into four groups or phases (Table I): the first is volatile synthetic organic chemicals; these are chemicals like trichlorethylene and tetrachloroethylene which are most frequently found in groundwaters. The first data on those substances appeared in 1978/79, and the list has been growing since then.

In the second phase, we will be dealing with the greatest number of substances and the greatest variety. The organics are primarily pesticides and other synthetic organic chemicals, often just discovered in the last years as potential contaminants in groundwater. We are examining all of the old inorganics and considering additional ones - and then taking a comprehensive new look at the microbial contaminants going far beyond just coliforms and turbidity.

Phase 3 is a new look at the radionuclides and that will occur beginning next year. Phase 4 is a comprehensive reexamination of disinfection byproducts starting with the THMs, but, of course, going far beyond because a tremendous amount of new information has been developed since the 1979 THM interim regulation was issued.

TABLE I.

Regulatory Framework

Phase I	**VOCs**
Phase II	**Organics/Pesticides, Inorganics, Microbials**
Phase IIA	**Fluoride**
Phase III	**Radionuclides**
Phase IV	**Disinfection By-Products (e.g. THMs)**

PHASE I REGULATIONS

These 10 chemicals (Table II), the volatile synthetic organics (VOCs), were selected as our evaluation group because of frequent detections in groundwaters. They are among the most common synthetic chemicals that are manufactured, many of them in millions of pounds per year quantities in the United States alone, relatively inert substances, and fortunately, mostly relatively non-toxic except for benzene, for example. That is one of the reasons that they are so widely used, because they do not have immediate toxicology associated with them in most cases: but there are a few exceptions.

In order to improve our understanding of the true occurrence of those substances in drinking waters in the United States, particularly in groundwaters, we completed a major national survey of 1000 groundwater supplies in two groups, and half of the group - approximately 500 was randomly selected (Table III). On frequencies, as one would have expected, tetrachloroethylene, trichloroethylene, 1,1,1-trichloroethane and the dichloroethylenes were found most frequently. Approximately 25 percent of the samples that were collected randomly were positive for one or more of these substances.

The detection limits were on the order of two tenths of a part per billion. Approximately 90 percent of those positives were very close to the detection limit, less than one part per billion. My suspicion is that many of those 90

TABLE II.

Phase I : VOCS

Trichloroethylene	Vinyl Chloride
Tetrachloroethylene	Benzene
Carbon Tetrachloride	1, 1-Dichloroethylene
1, 1, 1-Trichloroethane	Dichloromethane
1, 2-Dichloroethane	Chlorobenzenes

TABLE III.

Summary of GWSS Occurrence Data

RANDOM SAMPLE
n=466

Parameter	Frequency %	Maximum ug/L
TETRACHLOROETHYLENE	7.3	23
TRICHLOROETHYLENE	6.4	78
1,1,1-TRICHLOROETHANE	5.8	18
1,1-DICHLOROETHANE	3.9	3.2
1,2-DICHLOROETHYLENES (cis and/or trans)	3.4	2.0
CARBON TETRACHLORIDE	3.2	16
1,1-DICHLOROETHYLENE	1.9	6.3
m-XYLENE	1.7	1.5
o-+p-XYLENE	1.7	0.9
TOLUENE	1.3	2.9
1,2-DICHLOROPROPANE	1.3	21
p-DICHLOROBENZENE	1.1	1.3
BROMOBENZENE	0.9	5.8
ETHYLBENZENE	0.6	1.1
BENZENE	0.6	15
1,2-DICHLOROETHANE	0.6	1.0
VINYL CHLORIDE	0.2	1.1
DBCP	0.2	5.5

percent that were close to the detection limit were either the leading edge or the trailing edge of a plume of contamination, but more probably, the result of atmospheric fallout being brought down in the rain and migrating into relatively shallow aquifers. The average levels usually were about one part per billion or less across the board. Only about one percent of the random group exceeded 5 parts per billion.

Obviously, there are cases where one would find considerably higher concentrations, but in general, at these levels exposures from sources other than drinking water usually are by far much more significant in terms of daily body burden. As an example: benzene, one of the major components of unleaded gasoline is certainly found in groundwaters as perhaps a tracer in cases of leaking underground storage tank occurrence. But still, except in very unusual cases, inhalation of air is going to give a much greater dose than the few micrograms that one might get occasionally in drinking water. We have to keep that perspective in mind - which however, does not say we are not concerned about drinking water.

Early next year, our first proposal of recommended maximum contaminant levels for nine of these substances is going to be published. That will trigger the major regulatory actions toward eventually arriving at specific maximum contaminant levels for these substances in drinking water.

PHASE II REGULATIONS

The second group of substances that is being examined for revised regulations consists of the synthetic organic pesticides, the broad spectrum of inorganic chemicals and the biological contaminants (Table IV). Among the inorganics (Table V), this first group is the group that is already regulated in the National Interim Primary Regulations, and in many cases the numbers are going to change from what they were originally.

In addition to those, we are re-examining a much longer list of chemicals, including such well-known substances as asbestos and sodium and others like copper that is a corrosion product in many distribution systems, and aluminum which is one of the major substances used in the treatment of drinking water. I think the items of most concern in this group of substances are those that are produced as a result of corrosion where water in contact with piping leaches out metals such as lead, copper, perhaps zinc and other substances that are present in distribution systems.

Among the synthetic organics that we are examining for possible regulation (Table VI) there are six that are already regulated in the interims - all pesticides. But in the second group that is being examined, about 75 percent of these are pesticides, and they include such items as ethylene dibromide, dibromochloropropane, Aldicarb and many other pesticides, which in recent times are being found in areas where they were applied for agricultural use. We will not necessarily regulate all of these substances. We are concerned about pesticides that are migrating to water supplies, and about pesticides that are used as part of the

TABLE IV. **PHASE II**

- **SYNTHETIC ORGANIC CHEMICALS/ PESTICIDES**
- **INORGANIC CHEMICALS**
- **MICROBIOLOGICAL CONTAMINANTS**

TABLE V.

PHASE II : INORGANIC CHEMICALS

- NIPDWR:

Arsenic	Mercury
Barium	Nitrate
Cadium	Selenium
Chromium	Silver
Lead	Fluoride

- Other IOCs:

Aluminum	Vanadium	Berylium
Antimony	Sodium	Thalium
Molybdenum	Nickel	Cyanide
Asbestos	Zinc	
Sulfate	Corrosion	
Copper		

control of waterborne pests, either in water supplies themselves or in the watersheds.

Among the biological contaminants (Table VII) in addition to the original existing standards of coliforms and turbidity which have served very well, we know a lot more about biological contamination of drinking water than was known in 1912 or so when the first standards were proposed. In the United States, in about the last ten years, there were at least 20,000 cases of waterborne giardiasis that have been identified. In 1980 alone, there were somewhere between 20,000 and 100,000 cases of waterborne diseases of various sorts. Most of them go unreported. So I do not think there should be any complacency about the fact that biological contamination in many places, because of inadequate treatment or inadequate source protection, is still a very significant visible risk from drinking water consumption.

Viruses are the perennial concern and again, a matter of inadequate treatment or inadequate source control in some situations. From various monitoring that has been done, it appears that legionella are very common organisms inhabitating water distribution systems and home plumbing systems, which is not to say that ingestion of water-contagious legionella causes Legionnaire's Disease. But it does appear that the root of infection may be from inhalation of the organisms by susceptible individuals.

TABLE VI.

PHASE II : SYNTHETIC ORGANIC CHEMICALS (SOCs) / PESTICIDES

- NIPDWR

Endrin	2,4-D
Lindane	2,4,5-TP (Silvex)
Methoxychlor	
Toxaphene	

- OTHER SOCs

Chlordane	Aldicarb	Glyphosate
Pentachlorophenol	DBCP	Carbofuran
Atrazine	Simazine	Epichlorohydrin
PAHs	PCBs	Formaldehyde
Phthalates	Adipates	Endothal
Acrylamide	Dinoseb	Heptachlor
Butachlor	Alachlor	Ethylbenzene
Dalapon	Diquat	Freons
Picloram	Endothall	Trichlorobenzene
1,2-Dichloropropane	Vydate	Ethylene Dibromide
2,3,7,8-TCDD (Dioxin)	Toluene	1,1,2-TCE
Acrylonitrile	Xylene	1,1-Dichloroethane
Hexachlorobenzene	Styrene	Hexachlorocyclopentadine

TABLE VII.

PHASE II: MICROBIALS

- **NIPDWR**
  - COLIFORMS
  - TURBIDITY
- **OTHER MICROBIALS**
  - GIARDIA LAMBLIA — TREATMENT REQUIREMENT
  - STANDARD PLATE COUNT
  - VIRUSES
  - LEGIONELLA

Here is a case where we may be dealing in a hybrid where this may be a contaminant in drinking water but the transport of that from the water to the air and the inhalation may be the cause of a problem. The same thing may hold for VOCs and may hold for one of the next group, radon, which is a gaseous radionuclide found naturally in certain areas of the country. In addition to those radionuclides that already have been regulated, we are seriously looking at natural uranium and natural radon. (Table VIII).

Phase 4 (Table IX) is a comprehensive re-examination of all disinfection-related products, starting with the trihalomethanes but going much beyond them. There is a much greater understanding now about the byproducts of disinfection, the organics, but also the disinfectants themselves, like chlorine and chlorine dioxide that are used in some situations. These that are deliberately added to the water almost invariably are present in much greater quantities than any inadvertant contaminant that one would ever find in a drinking water supply; they not only contact the largest population exposed, but also likely are present in the highest concentrations.

TABLE VIII.

PHASE III: RADIONUCLIDES

- **NIPDWR**
  - **RADIUM 226 & 228**
  - **GROSS ALPHA**
  - **BETA PARTICLE/PHOTON RADIOACTIVITY**
- **OTHERS**
  - **URANIUM**
  - **RADON**

TABLE IX.

PHASE IV: DISINFECTANT BY-PRODUCTS

- **NIPDWR**
  TRIHALOMETHANES (THMs)
- **BY-PRODUCTS OF OTHER DISINFECTANTS**
  CHLORAMINES
  CHLORINE DIOXIDE
  OZONE

COMPLIANCE WITH REGULATIONS

Table X provides an idea of some of the compliance results in the United States since the implementation of the Safe Drinking Water Act interim regulations. There are about 59,000 public water systems in the United States and they serve somewhere between 180 and 200 million people; 3,000 of those exceed 10,000 persons in population; the other 56,000 are on the low end, they serve less than ten percent of the population.

About 50,000 of the 59,000 have not reported violations in a recent count. There are still about 1,000 persistent biological contaminant violations even when all it takes is chlorination to prevent it from happening. This is a small community system problem. There is still intermittent contamination occasionally reported and still turbidity problems in surface water supplies that are not filtering. There were about 500 violations of radionuclide standards - all natural radium. There were about 1,500 to 3,000 inorganic standards violations, half of these are fluoride, the others are odds and ends - a few arsenic, a few selenium, and so forth.

The water quality situation is better now than it was ten years ago before there was a national regulation. However, there isn't any magic solution that occurs. There still can be non-compliance. The approach is to continually move forward toward getting the maximum possible number of communities having the best quality water that they can reasonably have.

TABLE X.

NIPDWR MCL COMPLIANCE

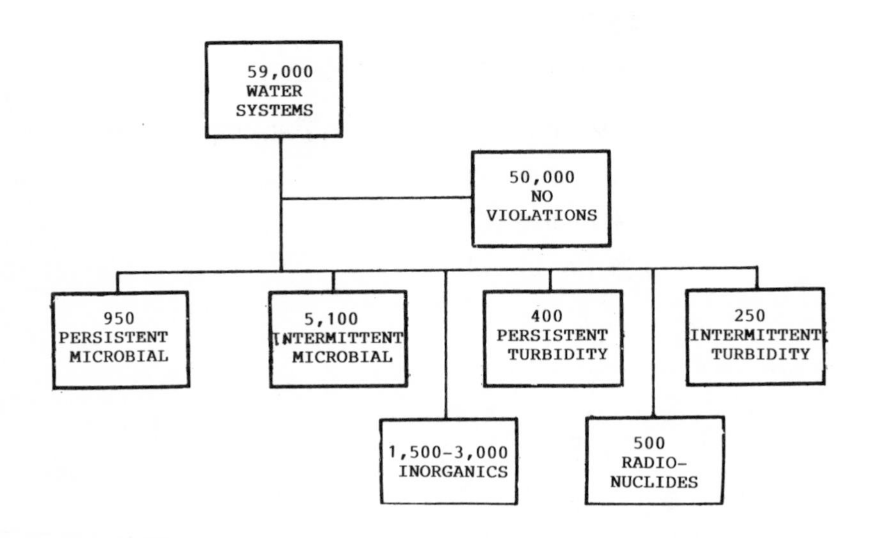

CONCLUSION

In summary, what I would like to leave with you is that there are a number of potential contaminants in drinking water that are of concern. Contrary to what we believed on the passage of typhoid and cholera and other waterborne diseases in North America - biological problems still do exist. There is no reason for them to exist because we know how to deal with them. We know how to filter water and disinfect water. The problem is making it happen in those places where these treatments do not happen, mostly the small communities. There is still particular concern with corrosion-related products and with disinfection byproducts - those are the substances where one has the largest population exposure potential and often highest concentrations. There is particular concern, especially recently, about groundwater contamination, especially pesticides and some of the synthetic organic chemicals that may be found. However keep in mind, there are many other routes of exposure to most of those substances besides drinking water, and it is important to control them too.

People that are in the water industry should present themselves as people who are aggressively trying to stop these things from happening, aggressively trying to control contamination where it exists, prevent it from occurring, and treat it when it does occur.

We all agree that it is intrinsically good to have clean drinking water. It is essential; it is a birthright maybe! I also do not believe that we need to justify providing clean safe drinking water on the premise that drinking water contamination is significantly related to cancer risks in the general population; it is extremely unlikely to be the case, much as Bruce Ames recently has pointed out. But it is important that drinking water be clean and safe and that we as a society decide that we are going to make it as clean and safe as we can possibly make it - because there is no need to tolerate less, and the cost is reasonable.

There are three areas that need to be emphasized: preventing contamination of drinking water sources; providing treatment when it is needed, that is tailored to the type of contamination; and of course, monitoring to ensure that we know what is in the water so that we can take appropriate action when contamination is found and needs to be dealt with.

CHAPTER 15

FEDERAL PROTECTION OF GROUNDWATER

Timothy L. Harker, Esquire

Kadison, Pfaelzer, Woodard, Quinn, & Rossi
2000 Pennsylvania Avenue, N.W.
Washington, DC 20006

An Office of Management and Budget policy official during the Carter Administration referred to groundwater protection as a "sleeping giant" of environmental issues. The President's Council on Environmental Quality at that time indicated that groundwater protection would be a priority environmental consideration in the 1980's. Developments of the last few years demonstrate the accuracy of these predictions. Having gone from litterbugs in the 1960s to Love Canal in the 1970s, America has focused on groundwater and related public health issues as its biggest environmental cause celebre. The sequence perhaps, is, logical. With the regulatory focus of the Environmental Protection Agency shifting increasingly in the direction of environmental health and protection of the food supply, it makes sense for EPA to be concerned with whether groundwater, a major source of food in the United States, is safe.

Many industries and municipalities utilize groundwater as a resource, either as a food (for drinking and food processing) or in production and other commercial activities. Many industries and municipalities also may adversely affect groundwater through waste disposal, and through unplanned or inadequate commercial and residential development. Both those who use and those who affect groundwater adversely have a major stake in the current development of a national groundwater protection strategy. These are the concerns of the food and agricultural community, including those who supply our nation's drinking water as well as the concerns of major segments of American commercial, residential and industrial development, including oil and gas industries, the chemical industry, mining, housing development, and municipal government. Some of these concerns are as follows:

A groundwater protection strategy at the federal level must account for the potential conflict between those who may abuse groundwater. In some cases the same entity may do both, e.g., municipal landfill and public water supply activities may conflict. Hence, a groundwater protection strategy must be based on a system of priorities that takes account of the economic importance not only of groundwater, but of those who may adversely affect it, i.e., not all groundwater is worth the cost of stringent protection in light of its current quality and in view of the economic and social costs of providing stringent protection. Not all groundwater is important as a source of drinking water, given available alternatives.

A federal groundwater protection strategy also must determine not only the significant sources of groundwater, but the significant sources of impact on groundwater and establish priorities accordingly. By failing to establish such priorities, we may fail to protect the most important uses and sources of groundwater. For example, if septic tanks and urban and storm water runoff (both of which occur near where groundwater is used for drinking) are largely uncontrolled, rigid federal regulations of other sources of groundwater contamination (such as waste disposal) may be relatively ineffective. It may be more cost-effective to focus on land use control by state and local government directed at protection of important recharge zones, rather than focusing heavily on discharge limitations. In particular, discharge restrictions in areas that may be located under or near groundwater but that are not important recharge areas may not be of great protective value for an aquifer if important recharge areas are ignored. Stringent federal limitations without land use controls of commercial and residential development in recharge zones may be nonproductive. In light of the conerns of those who utilize and those who impact groundwater, the current and projected federal program for groundwater protection is described below.

The Comprehensive Environmental Response, Compensation and Liability Act ("CERCLA" or the "Superfund Act") was enacted in December, 1980 to fill what was widely considered to be a large gap in federal environmental protection statutes. It is aimed principally at cleaning up thousands of abandoned hazardous waste disposal sites. The salient environmental feature of most of these sites is their threat to groundwater. Response actions taken by states and EPA at such sites frequently involve very costly groundwater monitoring and remedial engineering methods (grouting, slurry wall, leachate collection, carbon filtration of contaminated drinking water) designed to protect groundwater. EPA has authority to provide alternate water supplies to communities with contamination of private wells or the central system. The cost of groundwater protection efforts under the Superfund Act undoubtedly will run into billions of dollars.

Under the Resource Conservation and Recovery Act ("RCRA"), EPA has adopted hazardous waste and solid waste disposal regulations that in significant measure are aimed at groundwater protection. EPA will utilize siting criteria for new hazardous waste disposal facilities and an endangerment standard as reflected in numerical drinking water standards to determine the adequacy of hazardous waste disposal. EPA has adopted additional technological standards and monitoring requirements for hazardous waste disposal aimed at groundwater protection. In addition, federal criteria defining "open dumps" utilize groundwater protection as a principal objective to be used by the states in determining acceptable solid waste disposal practices.

Under the Safe Drinking Water Act, EPA protects groundwater through Underground Injection Control (UIC) regulations and through "sole source" aquifer protection. The UIC portion of that Act was the first substantial federal legislation directed solely at groundwater protection. EPA in its RCRA regulations borrowed the most salient feature of the UIC program, i.e., the use of the criterion, "endangerment of drinking water sources." The UIC program will include siting criteria, technological standards and monitoring requirements aimed at groundwater protection. The sole source aquifer program under the Safe Drinking Water Act is aimed at protecting the recharge zones of designated aquifers from adverse federally-sponsored or funded development.

The Clean Water Act traditionally has been viewed as protecting surface waters used in interstate commerce. While EPA may take account of the relationship between surface and groundwater in granting permits under the NPDES portion of this Act, this authority is unimportant as a source of groundwater protection.

Other federal statutes which have less bearing on groundwater protection include the Federal Surface Mining Control and Reclamation Act, the Toxic Substances Control Act (TSCA), and the Federal Insecticide, Fungicide, and Rodenticide Act (FIFRA). The Surface Mining Act provides siting criteria and reclamation requirements that include the objective of groundwater protection through mine drainage control and preservation of recharge zones.

The Toxic Substances Control Act includes the authority to regulate the disposal of any chemical. To date, EPA has done so with respect to PCBs and the highly toxic dioxin, TCDD. EPA is considering adoption of regulations for leaking underground storage tanks (LUST) under TSCA.

FIFRA provides authority for label restrictions on pesticide use. While this may be used to protect groundwater from pesticide contamination, EPA has not done so to any great extent. Most label restrictions on pesticide use apply to surface water, and generally to public water supplies or wells. Recently, however, a registrant for the pesticide, Aldicarb, agreed to terminate its use on certain crops in Long Island (Suffolk County) because the pesticide was leaching from the fields to groundwater. Yet, the relationship between pesticide application and groundwater, particularly due to irrigation return flows, to my knowledge, does not appear to have been explored by EPA.

With the exception of the Superfund Act and UIC program under the Safe Drinking Water Act, none of these federal statutes is directed primarily at groundwater protection. This has led some critics to contend that the current federal approach to groundwater protection is fragmented and contains major gaps. Critics of the current approach have called for a comprehensive program based on:

1) non-degradation of groundwater
2) the adoption of groundwater quality standards and effluent limitation
3) the implementation of land use controls aimed at protecting recharge zones.

However, other critics of the current federal approach contend that it is already misdirected, and that a system should be established that ranks aquifers according to their social and economic importance, and that determines through a cost/benefit approach which groundwater should be protected to the high standards of drinking water quality.

CHAPTER 16

THE WORLD HEALTH ORGANIZATION AND GUIDELINES AND EUROPEAN ECONOMIC COMMUNITY DIRECTIVES

J.R. Hickman, B. Pharm., B.S.

Department of National Health and Welfare
Health Protection Branch
Ottawa, K1A 0L2
Canada

INTRODUCTION

In discussing regulatory aspects related to drinking water and chemicals, I propose to deal mainly with the question of drinking water standards and guidelines; in particular, I have been invited to discuss the World Health Organization's (WHO) new guidelines for drinking water quality and the Directive of the Council of the European Communities of July 15, 1980, relating to the quality of water intended for human consumption.

For the purposes of this paper, I shall use the term "standards" as defined by Whyte and Burton (1980), viz:

> "Standards are prescribed levels, quantities or values which are regarded as authoritative measures of what is a safe enough, or acceptable, amount of . . . contamination or exposure to risk".

Depending on the legal framework which exists (this varies from country to country), standards usually are codified in regulations and they have the power of legal enforcement behind them. Guidelines, on the other hand, are more in the nature of recommended levels and do not have the backing of the law to ensure compliance with them.

THE ROLE OF WHO IN REGULATION OF DRINKING WATER QUALITY

WHO first published its International Standards for Drinking Water in 1958. The purpose was two-fold: (1) as part of the International Sanitary Regulations to govern the quality of drinking water supplied at ports and airports, and (2) to provide a basis for the formulation of national drinking water standards. Revised editions were published in 1963 and 1971. In addition, the WHO published separate European Drinking Water Standards, the last edition being published in 1970. The purpose of these European standards was "to encourage countries of advanced economic and technological capability to attain higher standards than the minimal ones specified in the International Standards for Drinking Water." It also was noted that industrial development and intensive agricultural practices in Europe created hazards to water quality.

The approach taken in the new Guidelines for Drinking Water Quality, as the document is titled, recognizes that the possibility for providing safe water varies greatly in different areas of the world. It also recognizes that a large part of the world's population does not have access to supplies piped into individual homes but must rely on community wells or standpipes, or on bottled water supplies. It recognizes that treatment technologies that are feasible in advanced countries may be quite impracticable in developing countries, and that supplies for rural villages may require different considerations from those for large metropolitan areas. Each case will require separate considerations, and the standards that are appropriate may differ appreciably.

By providing a common basis for considering health aspects associated with contaminants in drinking water, it is intended that potential health risks will be identified and receive adequate consideration along with technological and economic feasibility in national decisions relating to drinking water supply.

WORKING ARRANGEMENTS

Unlike previous revisions to the WHO International Standards for Drinking Water where the short-term assistance of consultants had been used to work with the WHO secretariat, the new guidelines represent a major effort on the part of both WHO and hundreds of scientists and technologists from thirty countries (Argentina, Australia, Belgium, Bulgaria, Canada, Czechoslovakia, Denmark, Egypt, France, Ghana, Greece, Federal Republic of Germany, Hungary, India, Israel, Japan, Luxembourg, Malaysia, Mexico, Netherlands, Nigeria, Senegal, Sudan, Sweden, Switzerland, Thailand, United Kingdom, USA, and USSR).

Expert groups were convened to develop criteria and recommend guideline values for microbiological, biological, chemical (inorganic, organic), organoleptic and radiological parameters. A separate task group dealt with the application of the guidelines, especially in developing countries, and with monitoring and surveillance activities.

Reports of all of these expert groups were reviewed by the National Focal Points for the WHO Environmental Health Criteria Programme; this provided an opportunity for input and comment from individual countries which eventually would become the "clients" for the final product. Finally, in March 1982, a group of scientists, water technologists and administrators from sixteen countries and several international agencies reviewed the documents to ensure that input and comments from National Focal Points had been taken into account properly, and to ensure consistency in approach on the part of the nine individual expert groups which had participated.

The new WHO Guidelines are being published in three volumes. Each of these serves a different purpose. Volume I, already published, is a compendium of the guideline values with a brief rationale used in deriving the recommended values; information also will be provided on their practical application (e.g., remedial actions, monitoring, etc.). It is expected that Volume I will be useful to waterworks operators, public health officials and administrators responsible for the provision of safe drinking water supplies.

Volume II, to be published shortly, is a compilation of the scientific criteria used by the various expert groups in deriving the guideline values. It elaborates on the scientific aspects, especially the toxicological and epidemiological evidence available to the various task groups.

Volume III is intended mainly for the lesser developed countries. In recognition of the fact that sophisticated approaches dependent on advanced technology may be the final goal, it introduces concepts, standards and procedures that can be adapted to local conditions to produce good interim results and a consequent drop in water-borne disease for communities embarking upon a drinking water quality control programme.

NATURE OF WHO GUIDELINE VALUES

At the risk of redundancy, I must stress the nature of the guideline values. They are intended to assist the appropriate agencies in different countries to develop standards. They are not, in themselves, standards.

The definitions of guideline values need, therefore, to be carefully considered. These are:

(a) A guideline value represents the level (a concentration or a number) of a constituent which ensures an aesthetically pleasing water and does not result in any significant risk to the health of the consumer.

(b) The quality of water defined by the Guidelines for Drinking Water Quality is such that it is suitable for human consumption and for all usually domestic purposes, including personal hygiene. However, water of a higher quality may be required for some special purposes, such as renal dialysis.

(c) When a guideline value is exceeded, this should be a signal: (i) to investigate the cause with a view to taking remedial action; (ii) to consult with responsible public health authorities for advice.

(d) Although the guideline values describe a quality of water acceptable for lifelong consumption, the establishment of these guidelines should not be regarded as implying that the quality of drinking water may be degraded to the recommended level. Indeed, a continuous effort should be made to maintain drinking water quality at the highest possible level.

(e) The specified guideline values have been derived to safeguard health on the basis of lifelong consumption. Short-term exposures to higher levels of chemical constituents, such as might occur following accidental contamination, may be tolerated but need to be assessed case-by-case, taking into account, for example, the acute toxicity of the substance involved.

(f) Short-term deviations above the guideline values do not necessarily mean that the water is unsuitable for consumption. The amount by which, and the period for which, any guideline value can be exceeded without affecting public health depends on the specific substance involved.

It is recommended that, when a guideline value is exceeded, the surveillance agency (usually the authority responsible for public health) should be consulted for advice on suitable action, taking into account the intake of the substance from sources other than drinking water (for chemical constituents), the likelihood of adverse effects, the practicality of remedial measures, and similar factors.

(g) In developing national drinking water standards based on these Guidelines, it will be necessary to take account of a variety of local geographical, socioeconomic, dietary and industrial conditions. This may lead to national standards that differ appreciably from the guideline values.

(h) In the case of radioactive substances, the term "guideline value" is used in the sense of "reference level" as defined by the International Commission on Radiological Protection (ICRP).

Certain assumptions are inherent in the guideline values. One of these is that the daily per capita intake of water is two liters. This probably slightly overestimates the average daily consumption in temperate zones. In Canada, for example, surveys show that the average daily consumption of tapwater and tapwater-based beverages is 1.34 liters (an amount slightly greater than in the U.K. or Netherlands where similar surveys have been made). Total daily fluid intake in Canada is about 2 liters, the remainder being imbibed as milk, alcoholic beverages and soft drinks (Health and Welfare Canada, 1981). On the other hand, it should also be recognized that a significant number of individuals consume four liters or more per day.

Those living in hot regions, especially under desert conditions, may consume much more (Molnar, 1946, Galagan et al., 1957). I should also point out that the consumption per unit body weight basis is much higher in children. In Canada, the five-and-under age group consumes about twice as much tapwater per kg body weight as adults. Given the particular sensitivity of children to such water contaminants as lead and nitrate, such considerations may be significant.

Another matter that deserves particular mention is the extrapolation approach used to derive the guideline values for organic chemicals that are known carcinogens. The guidelines note that the methodology adopted to arrive at guideline values is very different in the case of these substances, in that it relies upon a very conservative hypothetical mathematical model that cannot be experimentally verified. The uncertainties involved are significant, and at least of the order of about two orders of magnitude (i.e., from 1/10th to 10 times the recommended number).

Since these calculations were made in 1980, there have been considerable developments in the field of carcinogenic risk assessment (Krewski, 1983), but these in no way increase confidence in the guideline values that were derived. The problems inherent in assessing the risks associated with the trace quantities of organics were well recognized by the Task Group convened by WHO in 1980 to recommend guideline values for organics, and it is noteworthy that the Task Group recommended that WHO should plan to re-evaluate these values within five years in view of the complexities of the issue and the rapid advances in risk assessment methodology that were taking place.

In the case of some of the organic compounds considered, it was recognized that their most important influence on the quality of drinking water was in relation to aesthetic and organoleptic aspects rather than health effects. Such substances often could render a water completely undrinkable at levels well below any that would give rise to concern from a health aspect.

DIRECTIVE OF THE EUROPEAN COMMUNITIES RELATING TO THE QUALITY OF WATER FOR HUMAN CONSUMPTION

In contrast to the WHO Guidelines, which are advisory in nature and provide a scientific basis for countries to develop their own standards, the European Community Directive of July 15, 1980, relating to the quality of water intended for human consumption establishes finite standards to be respected by Member States. As noted by Amavis and Smeets (1981), adoption of this directive signifies agreement between 9 countries on 62 parameters, their numerical values and their monitoring.

It is relevant to note that the EEC previously issued another Directive in 1975 titled "The Quality Required for Surface Water Intended for the Abstraction of Drinking Water in the Member States. Briefly, this Directive gives the quality requirements which surface water used for preparing drinking water must meet. Three categories of water (as defined by physical, chemical and microbiological characteristics) are recognized depending upon the degree of treatment used. Thus, the EEC already has moved some distance toward the recommendations of OECD (1982) which require a system of matching instruments that are independent but harmonized (Figure 1). The OECD report notes that, despite good available experience, a number of its Member countries still only use some elements of the system, and often without the necessary interconnections.

COMPARISON OF WHO GUIDELINES AND EEC STANDARDS

There is a remarkable similarity between the WHO and EEC lists in terms of health-related inorganic substances that are included and the guideline values (WHO) and maximum admissible concentrations (MAC)(EEC). The European standards set levels for silver and nitrite, both of which were considered in the WHO document but for which guideline values were not set. The value for cyanide set by WHO (0.1 mg/L) is higher than that of the EEC. In the case of lead, both documents specify 0.05 mg/L, but the EEC makes exceptions to permit up to 0.1 mg/L when lead pipes are present.

The WHO guidelines are notable for the range of organic substances for which guideline values have been set. These include a number of chloroalkenes and chloroalkanes, chlorophenols and chlorobenzenes. Notable is a guideline value for chloroform of 30 microg/L. In contrast, the EEC standard has no MAC for organics other than pesticides. In the case of haloforms, a "Guide Level" of 1 microg/L is given (with the comment that haloform concentrations must be as low as possible). The WHO document lists values for 7 commonly occurring pesticides individually. The EEC document requires that not more than 0.1 microg/L of any individual pesticide be present with a total not exceeding 0.5 microg/L. The EEC has retained the

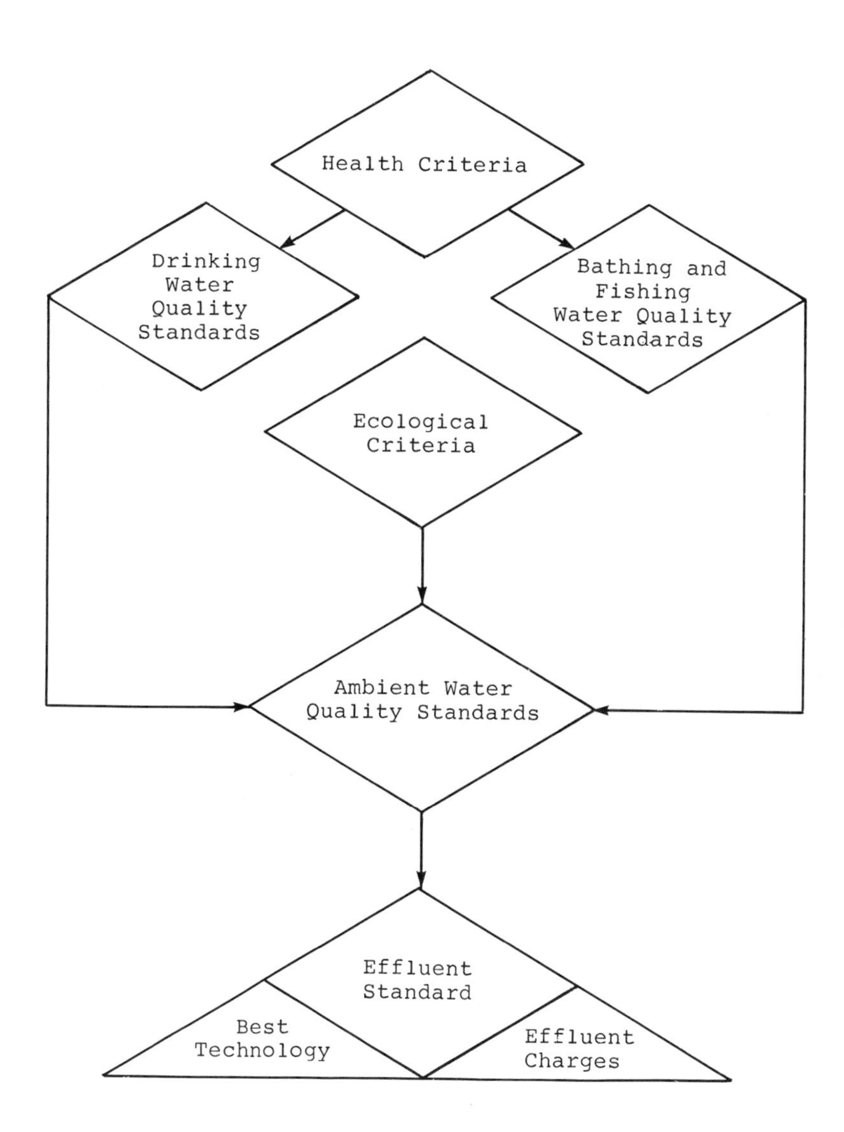

Figure 1. Dependence of water quality standards.

traditional PAH standard based on five reference substances. This was dropped in the WHO standard but replaced by a specific guideline value for benzo(a)pyrene.

Differences between the two documents are much more noticeable among those parameters affecting the organoleptic and aesthetic qualities of water. In particular, parameters such as iron and manganese tend to be more closely controlled in the European standard.

CONCLUSION

In conclusion, the new WHO guidelines provide a scientific basis for establishing standards with respect to health protection. By comparison, the EEC standards which were derived independently show remarkable similarities in terms of both parameters included and the numerical limits derived for them.

REFERENCES

Amavis, R. and Smeets, J. (1981), "Directive of the European Communities Relating to the Quality of Water for Human Consumption," Sci. Total Environ. 18:293-300.

Galagan, D.J. et al. (1957), "Climate and Fluid Intake," Publ. Health. Rep. 74:484.

Health and Welfare Canada (1981), "Tapwater Consumption in Canada," Environmental Health Directorate Report Series, 82-EHD-80.

Krewski, D. (1983), "Carcinogenic Risk Assessment, J. Amer. Statistical Assoc. 78:308-310.

Molner, G.W. et al, (1946) "A Comparative Study of Water, Salt and Heat Exchange of Men in Tropical and Desert Environments," Amer. J. Hyg. 44:411.

OCED (1982), "Control Policies for Specific Water Pollutants," (Paris, France; Organization for Economic Cooperation).

Whyte, A.V. and Burton I., (1980) Environmental Risk Assessment (New York, NY; John Wiley and Sons), p. 116.

CHAPTER 17

REGULATORY FLEXIBILITY AND CONSUMER OPTIONS UNDER THE SAFE DRINKING WATER ACT

Timothy L. Harker, Esquire

Kadison, Pfaelzer, Woodard, Quinn, & Rossi
2000 Pennsylvania Avenue, N.W.
Washington, DC 20006

Hazardous waste contamination of public and private water supplies, combined with financial instability of many small public treatment systems to comply with the Safe Drinking Water Act, appear to be a permanent part of our nation's public health concerns. State and local governments increasingly are confronted with the need to recommend alternatives to centralized treatment as sources of safe drinking water.

Bottled water is a principal alternative source for use in emergencies and on an interim or permanent basis for small public water supply systems. Accordingly, state and local governments need to learn more about the safety and the regulation of this alternate source, as well as its physical capability and legal capacity to serve as an alternative to centralized treatment under the Safe Drinking Water Act.

The following position paper, developed on behalf of the Drinking Water Research Foundation, addresses these issues.

THE FAILURE OF SMALL SYSTEMS TO COMPLY WITH THE SAFE DRINKING WATER ACT

The Safe Drinking Water Act of 1974 ("the Act") represents the first comprehensive national effort to assure safe drinking water. Under the direction of the Act, the Administrator of the Environmental Protection Agency has established Primary Drinking Water Regulations, including "Maximum Contaminant Levels" ("MCLs") which public drinking water systems must attain for such pollutants as coliforms, turbidity, inorganic chemicals, pesticides, radiation, and more recently, trihalomethanes ("THMs").

The promulgation of MCLs is not the end of drinking water problems, unfortunately. According to EPA, as of 1980, approximately 13,600 community water systems were in violation of one or more Interim Primary Drinking Water Regulations (1). The incidence of violation appears to be increasing. During fiscal year 1982 over 70,000 violations of the Interim Regulations were recorded by 20,000 community water systems. While most violations are of monitoring and reported standards, EPA estimates that more than 9,000 systems must improve treatment facilities in order to meet health-related drinking water standards, (48 Fed. Reg. 45505, Oct. 5, 1983).

There are approximately 60,000 community systems. Two-thirds (over 38,000) serve fewer than 500 people. EPA has concluded that compliance with the Interim Primary Drinking Water Regulations is a problem mostly for small systems, (serving less than 3,300 people). EPA reports that in 1982 the microbiological requirements were not met by 10% of the smaller systems, (48 Fed. Reg. 45505). Monitoring also indicates that between 1,500 and 3,000 community systems exceed current Maximum Contaminant Levels for certain inorganic contaminants, (48 Fed. Reg. 45504). Problems are primarily with arsenic, barium, lead (from pipe or solder corrosion) fluoride and nitrate.

The inability of small community systems to finance the improvements necessary to come into compliance with Interim Primary Drinking Water Regulations is an important cause of these high noncompliance rates. EPA estimates that of the community systems violating one or more of the MCLs, approximately 8,700 (63%) would require "relatively expensive treatment improvements" in order to achieve compliance. Approximately 3,300 of these systems would be "unable to make necessary improvements [in centralized treatment] using conventional local financing" (2). Of the community water systems unable to finance necessary improvements, 97% serve less than 2,500 consumers and 91% serve less than 500 consumers (2).

The statistics given above do not reflect compliance rates for the 163,000 "non-community systems" (systems having at least 15 service connections used by travelers or transients or regularly serving 25 or more non-resident individuals daily for at least 60 days per year) which are also subject to federal drinking water regulations. Non-community systems include motels, campgrounds, hospitals, and other water systems serving non-resident populations. The number of these systems in violation of drinking water regulations has not been estimated. EPA concludes that while compliance records for non-community systems are incomplete, these systems generally do not meet monitoring requirements on schedule. The significance of such frequent monitoring violations by public drinking water systems is clear:

Certain contaminants such as coliforms, turbidity and some inorganic and organic chemicals are widely detected in drinking water supplies and pose serious health risks when MCLs are exceeded. Without consistent or frequent oversight, these MCLs have a high potential for being exceeded. Such contaminants warrant national regulations with fixed minimum requirements, including regular monitoring requirements, (48 Fed. Reg. 45505).

The same factors responsible for non-compliance in small community systems often occur in non-community systems. The factors obstructing small system compliance with Primary Drinking Water Regulations include the following:

Small systems frequently are unable to take advantage of scale economics in treatment techniques, resulting in higher per-capita costs.

Small systems often lack full-time and properly trained system operators possessing adequate technical expertise.

Small systems often have less latitude in seeking alternative water supplies and therefore are more subject to local water quality problems than are larger systems, which can look further for better quality sources.

Some treatment techniques are simply economically unavailable to small systems with limited financial resources.

Local resistance to higher user fees can impede the ability of small systems to finance necessary improvements.

Nor do statistics on compliance failure on the part of small public systems reflect the estimated 5.5 million families in rural areas that are indicated in an EPA survey to be drinking water with toxic contaminants.

The non-compliance difficulties faced by small systems assuredly will be compounded in the near future, due to the potentially disproportionate burden which small systems will bear in complying with likely new standards for volatile organic compounds ("VOCs"), inorganic compounds and microbiological contaminants.

EPA is considering additional regulation of microbiological, organic and inorganic contaminants. Among options under evaluation for microbiological contaminants are additional criteria for parasites and viruses, disinfection as minimum treatment for all groundwater sources of public drinking water, and increased monitoring, (48 Fed. Reg. 45511). Among options under consideration for inorganic contaminants are more stringent MCLs for nitrate, arsenic, barium, cadmium and lead, control of which are "relatively expensive on a per capita basis for small public water systems" (48 Fed. Reg. 45504).

Additional regulations for inorganics also may involve sodium and corrosion control. EPA reports that "sodium is probably the most difficult substance to remove from drinking water" (48 Fed. Reg. 45516). EPA estimates that 16 percent of public water systems in the U.S. are highly aggressive and an additional 52% are moderately aggressive. Only a limited number of those public systems have instituted corrosion control. "Corrosion is a very significant concern" (48 Fed. Reg. 45517). It poses a serious economic impact and affects both the aesthetic quality and safety of drinking water.

Case-in-Point: VOCs and Trihalomethanes

VOCs are synthetic organic chemicals which EPA believes enter water supplies through improper disposal of hazardous wastes and industrial activities. These chemicals are not rapidly biodegraded nor readily adsorbed by soil particles; once they enter groundwater, they may remain for years or decades. EPA is concerned with VOCs in drinking water because of their possible effects on human health, as a possible cause of cancer and other diseases. Recent EPA surveys detected VOCs in about 28% of public water systems using groundwaters serving over 10,000 people, and in about 17% of such systems serving less than 10,000. Although the amount of VOCs detected in most systems was not high enough for EPA to consider them as health threats, some systems had significant contamination.

Initial estimates of the preliminary costs for controlling VOCs in public water supplies suggest that the monthly per-household cost increases for such treatment will be greatest in small systems, regardless of the technology employed, as shown in Table I.

EPA has addressed the problem of disproportionate costs in the case of a different set of contaminants, the THMs. The EPA solution to this problem was simply to exempt all systems serving under 10,000 persons from the THM standard (3). Such systems need not provide the level of drinking water safety prescribed for larger systems.

PROPOSAL: BOTTLED WATER AS A MEANS OF ACCOMPLISHING SMALL-SYSTEM COMPLIANCE WITH PRIMARY DRINKING WATER REGULATIONS

Bottled water is receiving increasing attention as an alternative means for small systems to overcome water quality problems. The provision of bottled water to serve the drinking and cooking water requirements of consumers is an obvious alternative to centralized treatment designed to bring all tap water up to the quality needed for the less than one percent that is used for drinking and cooking. Much of the benefit of centralized treatment is simply flushed away.

Bottled water is capable of serving at least four different roles for small systems facing non-compliance problems.

TABLE I. SUMMARY OF INCREASED PER-HOUSEHOLD COSTS FOR REDUCING TRICHLOROETHYLENE (a VOC) LEVELS IN DRINKING WATER

	Population Served		
Technology	100-499	1,000-2,499	10,000-24,999
A. 90% Removal (500 mg/L to 50 mg/L			
Aeration (Packed Tower)	$4.65-6.49	$2.04-2.58	$0.72-0.92
Aeration (Diffused Air)	9.76	5.04	2.41
Adsorption (GAC)	13.48	7.62	2.03
B. 99% Removal (500 mg/L to 5 mg/L)			
Aeration (Packed Tower)	$5.11-7.43	$2.34-12.00	$0.85-1.13
Aeration (Diffused Air)	13.48	8.23	3.81
Adsorption (GAC)	13.95	7.87	6.32

Bottled Water As A Permanent Water Supply

First, bottled water can serve as a permanent supply of potable water for an entire small community or non-community system, or for residential areas served by private wells in an aquifer that has become contaminated. This option is most appropriate where alternatives, such as upgrading centralized treatment or extending service lines to a complying central system, are not available without incurring substantial capital costs, for which the small system or the community does not have the population base to provide funding. Available centralized treatment equipment may be obsolete or otherwise incapable of achieving MCLs without expensive upgrading. The system may be in violation of more than one treatment technology because no single treatment technique is effective with respect to all regulated contaminants.

Non-community systems also might use bottled water on a permanent basis. These systems face problems peculiar to their size, ownership, and the transient populations they serve. The principal focus for such systems is acute health effects, such as those arising from bacterial contamination, rather than chronic effects arising from long-term exposure. Many non-community systems are privately owned, and such owners often face an even more limited capacity to raise the capital necessary to achieve drinking water standards through centralized treatment.

Residential communities served by privately owned groundwater wells that have become contaminated also may use bottled water on a permanent basis (4).

The primary advantage of the bottled water option lies in its low capital intensity. Communities and non-community systems and residential areas serviced by contaminated private wells thus can achieve permanent compliance with all Primary Drinking Water Regulations without large capital expenditures. Not having to confront a substantial capital expenditure may be a significant inducement for a small system or for a community to come into compliance. For example, EPA indicates that some small systems remain unconvinced that the health benefits of large capital expenditures are warranted (48 Fed. Reg. 45504).

The use of bottled water where the financial capacity of the system is insufficient to support the capital expenditures necessary for upgrading centralized treatment would accomplish the purposes of the Safe Drinking Water Act considerably better than merely exempting small systems from various contaminant regulations, as was the approach taken with respect to THMs. The health considerations forming the basis of regulation in large systems are no less compelling for small systems.

Bottled Water As An Interim Solution

The second purpose which can be served by bottled water is as a temporary solution during the intermediate period while permanent centralized solutions are being devised. The installation of modern centralized treatment systems can be time-consuming, while the demand for water which meets applicable health standards is immediate. Even where bottled water is not the ultimate cost-effective permanent solution, its use in this interim capacity often is appropriate.

Such interim use is well precedented. For example, interim provision of bottled water was accomplished in Duluth, Minnesota in 1976 and 1977, while the public system was upgraded in order to remove unacceptable levels of asbestos. Similarly, the United States Government has consented to a stipulation agreement with the Alcova Acres Investment Company providing that the company will provide bottled water for cooking and drinking to all service connections served by the Alcova Acres Water System in Natrona County, Wyoming, until that system comes into compliance with MCLs for nitrates and selenium (5).

Bottled Water As An Emergency Alternative Source

A third use of bottled water by water systems facing water quality problems is to provide an emergency, short-term source of potable water to all consumers of a public system. Emergency situations of a temporary nature, when water quality problems raise acute health hazards, may create a pressing demand for potable water which can be met only by the rapid short-term importation of bottled water. This use of bottled water has been expressly recognized by the U.S. Army Corps of Engineers, the Federal Emergency Water Administration and the U.S. Environmental Protection Agency under the National Contingency Plan for responding to contamination of drinking water supplies.

The bottled water industry has considerable experience in supplying public drinking water needs during periods of emergency. In numerous emergency situations, from earthquakes in Los Angeles, to Hurricane Agnes, to the Johnstown floods, to the Three-Mile Island accident in Pennsylvania, to contamination of community wells from the release of hazardous materials, the bottled water industry has performed as an alternative to centralized treatment and to private wells.

Bottled Water As A Permanent Alternative Source For Special Segments Of Population

It is proposed that bottled water can serve a specialized purpose relating to particularly sensitive population groups. Some drinking water regulations are targeted at particularly sensitive population groups. For example, maximum nitrate levels are set to provide safety for small children and pregnant women, while higher levels are safe for other persons. Maximum safe levels for fluoride also may be higher for adults than for children. Recommended maximum sodium levels for persons on sodium-restricted diets need not apply for all others in the population. In some situations, in both small community systems and noncommunity systems, the most efficient means of providing the level of safety sought in the Safe Drinking Water Act is to permit the system to provide bottled drinking water to that portion of its consumers which is at risk from a particular contaminant, rather than requiring expensive centralized treatment in order to meet the specialized needs of an identifiable sub-group.

THE QUALITY OF BOTTLED WATER

The safety of water delivered by the bottled water industry should be more than adequate for small systems to meet the goals of the Safe Drinking Water Act. Bottled water is a highly regulated and monitored drinking water supply. All bottled water must: (a) come from protected sources and meet government safety and quality standards; (b) be bottled in facilities regulated as food plants; (c) be processed in accordance with practices approved by the federal goverment; (d) be delivered in bottles whose safety as food packaging is regulated by federal standards; and (e) be so labeled as to provide public notification whenever quality is substandard.

Bottled water is regulated as a food by the Federal Food and Drug Adminsistration ("FDA"). The FDA has issued quality standards (21 CFR § 103.35), and Good Manufacturing Practices (21 CFR § 129) for bottled water.

The quality standards have been broadened to include pesticides, mercury, radioactivity and trihalomethanes (applicable to all bottled water, unlike MCLs for THMs which apply only in public drinking water systems serving 10,000 people or more). FDA's standards are such that bottled water _must meet_ all applicable provisions of EPA's Interim Primary standards and all secondary drinking water criteria (which are only _guidelines_ for public water supplies).

FDA's Good Manufacturing Practices establish minimum criteria for buildings and facilities, raw materials, operating procedures, and controls. The Good Manufacturing Practices have been adjusted to provide for regular analytical testing of bottled water. Ozonation of bottled water, the principal source of disinfection, has been affirmed as Generally Recognized As Safe by the FDA. In addition to FDA, almost all states now regulate bottled water. All states in which the bottled water market is significant have comprehensive regulations, some of which go beyond FDA's standards.

Under provisions of the Safe Drinking Water Act, and in keeping with the 1978 Memorandum of Understanding between FDA and EPA, the EPA also has a significant impact on the regulation of bottled water. The Safe Drinking Water Act requires that FDA adopt revisions in EPA's national drinking water standards for bottled water, or publish in the _Federal Register_ reasons for not doing so (6).

FDA regulations provide control over food contact surfaces, including the equipment and packaging used in the processing and distribution of bottled water (7). In contrast, regulations for tap water under the Safe Drinking Water Act do not fully regulate delivery systems. Because many pipe networks are antiquated and deteriorating, water in compliance with applicable regulations when it leaves the treatment plant may not be in compliance when it arrives in the home. Bottled water is regulated with respect to packaging as well as content.

A final FDA safeguard requires public notification whenever the microbiological, physical, chemical or radiological quality of bottled water is substandard (8).

The bottled water industry has a comprehensive program of self-regulation. The International Bottled Water Association's ("IBWA") Plant Technical Manual provides members with detailed information on processing and quality control techniques to be used in the production and bottling of each industry product. IBWA employs an independent laboratory to audit the performance of each member company. This annual audit evaluates compliance with the association's performance requirements and FDA's safety regulations.

To keep the industry informed on the developments in quality control techniques and water processing and bottling, IBWA sponsors an on-going series of technical seminars and workshops open to members and non-members alike. IBWA provides an educational course and certification program for plant operators.

Frequency of inspection is an important measure of regulatory oversight. IBWA surveys of member companies since 1978 show that bottled water plants throughout the U.S. are inspected at least once a year by a federal, state, or local agency. Most plants are inspected several times per year.

LEGAL AUTHORITY FOR THE UTILIZATION OF BOTTLED WATER TO MEET PRIMARY DRINKING WATER REGULATIONS

The Safe Drinking Water Act, 42 U.S.C. § 300f-j, allows EPA to authorize the use of bottled water, where appropriate, to achieve the goals of the Act. The Act regulates the safety of public drinking water through the promulgation of "primary drinking water regulations" by the Administrator of EPA. Section 300f(1) of the statute defines "primary drinking water regulation" as a regulation which

> (A) applies to public water systems;
>
> (B) specifies contaminants which . . . may have an adverse effect on the health of persons;
>
> (C) specifies for each such contaminant either-
>
> (i) a maximum contaminant level . . . or
>
> (ii) . . . each treatment technique . . . which leads to a [sufficient reduction in the level of such contaminant . . .; and

> (D) contains criteria and procedures to assure a supply of drinking water which dependably complies with such maximum contaminant levels . . .

The term "maximum contaminant level" ("MCL") is defined in 300f(3) as the "maximum permissible level of a contaminant in water which is delivered to any user of a public water system."

Nothing in these definitions prohibits the Administrator from issuing regulations which permit small systems to comply with MCLs and treatment specifications by providing bottled water to all consumers on a permanent or temporary basis. The definition of MCL refers to contaminant levels in water "delivered to any user of a public water system"; delivery in bottles is not precluded.

Section 300g-1 of the Act, which states criteria for the selection of MCLs and treatment techniques, provides additional support for the authority of EPA to permit the use of bottled water to comply with the Act. Briefly, § 300g-1 requires the Administrator to select MCLs and required treatment techniques such that contaminant levels are reduced to levels which are as close to those which prevent adverse health effects "as is feasible." The term "feasible" is defined at § 300g-1(b)(3) as "feasible with the use of the best technology, treatment techniques, and _other means_, which the Administrator finds are generally available (taking costs into consideration)." (Emphasis added.) This section authorizes flexibility in the selection of policies to accomplish compliance with MCLs and technique requirements. Where such compliance is "feasible" and cost-effectively accomplished by means of bottled water, regulations permitting its use are authorized.

The legislative history of the Safe Drinking Water Act verifies this analysis. Senator Hart's extended discussion of the bill in the _Congressional Record_ explains the National Primary Drinking Water Standards as follows:

> "The standard for both national interim primary drinking water regulations and revised national primary drinking water regulations is that health shall be protected to the extent feasible. Obviously, as the determinations vary, the feasibility of a standard is governed not only by the cost of compliance but also with the extent of health risk; balancing between health risks on the one side and the cost of compliance on the other side must be weighed. Where health risks are great, higher costs may be incurred, perhaps even to the point of requiring alternative sources of drinking water" (9).

This weighing of health risks and compliance costs is particularly appropriate in the case of small systems facing difficult water quality problems. Under such circumstances, the provision of alternative sources may be the most cost-effective means, and perhaps the only practical means, of achieving the health protection goals of the Act.

The sections of the Act governing the granting of variances and exemptions from Primary Drinking Water Regulations verify the authority of the Administrator to permit systems to meet drinking water standards through the use of bottled water. The statute permits the granting of a variance to a public water system which, "because of characteristics of the raw water sources which are reasonably available to these systems, cannot meet the requirements respecting the maximum contaminant levels of such drinking water regulations despite application of the best technology, treatment techniques, or other means, which the Administrator finds are generally available (taking costs into consideration)" (10).

Under this language, the Administrator must find that "other means" of compliance with MCLs (such as bottled water) are <u>not</u> available before a variance may be granted.

Similarly, § 300g-5(a)(2) of the statute permits an extension in the compliance date for new public water systems "only if no reasonable alternative source of drinking water is available to such new system." Thus, the statute explicitly provides for the consideration and use of "alternative sources of drinking water" (presumably including bottled water) <u>before</u> an exemption may be granted, as a means of compliance with MCLs.

EPA has issued a policy statement discussing the special problems faced by small public systems in complying with the Act, and proposing small system strategies to mitigate acute health hazards and implement the most feasible solution to other types of violations as expeditiously as possible (11). The Agency's proposed small system compliance strategy explicitly authorizes an assessment of "alternative means of compliance" for small systems in violation of drinking water standards. "Among the alternatives to be considered are regionalization or some other inter-community arrangement, <u>alternative water sources</u>, blending or mixing of new and old sources, and, finally, treatment" (12).

OTHER GOVERNMENT POLICIES AUTHORIZING THE USE OF BOTTLED WATER TO MEET DRINKING WATER NEEDS

The utlization of bottled water during water supply emergencies not only is widespread in practice, but also is specifically authorized by several agencies bearing water-supply responsibilities.

EPA's National Contingency Plan Under The Superfund Act

The Comprehensive Environmental Response, Compensation, and Liability Act of 1980 (the "Superfund" Act) establishes broad Federal authority to respond to releases of hazardous substances, pollutants, and contaminants into the environment. Section 105 of the Act mandates revision of the National Contingency Plan originally published pursuant to § 311 of the Federal Water Pollution Control Act, 42 U.S.C. § 1321(c)(2), by requiring the inclusion of a section to be known as the "National Hazardous Substance Response Plan." EPA has adopted a detailed set of modifications to the National Contingency Plan in compliance with § 105 of the Superfund Act. Among the remedial actions explicitly authorized under this proposal is the provision of bottled water as a means of providing alternate water supplies when normal supplies are contaminated (13).

Department Of Interior's Emergency Water Supply Plan

The Department of Interior has promulgated an Emergency Water Supply Plan for the purpose of assuring an adequate supply and best use of water for national defense and essential civilian purposes. The Plan specifies that state and local governments should inventory the capability of bottled water companies in their areas to meet national emergency needs.

The U.S. Army Corps Of Engineers' Emergency Water Plan

The Army Corps of Engineers' authority to provide temporary supplies of clean drinking water during periods of emergency includes the procurement and distribution of bottled water (33 CFR § 214.10).

CONCLUSION

The cost of providing bottled water to consumers in a community or non-community system will vary greatly according to the particular circumstances of each case. Obviously, a small system considering the bottled water option will have to compare its cost with the cost of other available alternatives. What is crucial is that EPA and the states authorize use of the bottled water option by public systems so that, in those cases in which it proves to be the most economical option, those systems will be able to select it. It would be counterproductive to foreclose the use of bottled water as a matter of policy. In many cases, because of economic realities, the alternative to using bottled water is to tolerate the provision of drinking water which exceeds safety standards.

The bottled water industry is an experienced provider of safe drinking water to millions of consumers. It should play a role in accomplishing the purposes of the Safe Drinking Water Act, particularly as an alternate source to which small systems and small communities can turn for assistance.

REFERENCES

1. Environmental Protection Agency, "Small Systems Compliance for Public Water Supply Systems -- Safe Drinking Water Act" 45 Fed. Reg. 40222-40223 (June 13, 1980).

2. Environmental Protection Agency, "Small System Compliance for Public Water Supply Systems -- Safe Drinking Water Act" 45 Fed. Reg. 40223 (June 13, 1980).

3. 40 CFR § 141.12.

4. In re Exxon Company, U.S.A.; RCRA Docket No. 83-1027 and In re Mobil Oil Corporation; RERA Docket No. 83-1028; EPA Determination and Consent Order providing that companies shall provide bottled water to communities served by contaminanted private wells.

5. U.S. v. Alcova Investment Co., D.C. Wyo. No. C 82-030813, Stipulation for Entry of Judgement.

6. 21 U.S.C. § 349.

7. 21 CFR § 129.

8. 21 CFR § 103.

9. Congr. Rec., Nov. 26, 1974 at p.S20241.

10. 42 U.S.C. § 300g-4 (a)(1)(A).

11. 45 Fed. Reg. 40222 (June 13, 1980).

12. Idem at 40223. The statute authorizes the states to take over the regulation of public drinking water, provided that, inter alia, the state regulations are "no less stringent" than the federal requirements (42 USC 300g-2). The power of the states to permit public supplies to use bottled water thus will depend on whether the EPA is granted such discretion under the statute.

13. 40 CFR § 300.70(d); 47 Fed. Reg. 31180, 31218-19 (July 16, 1982).

CHAPTER 18

STRUCTURE AND REGULATION OF THE EUROPEAN BOTTLED WATER INDUSTRY

Rip G. Rice, Ph.D.

Rip G. Rice, Incorporated
Ashton, MD 20861

G. Wade Miller

Wade Miller Associates
Arlington, VA 22209

Abstract: In the United States, bottled water is an alternative option to municipal drinking water. This is because U.S. bottled water either is groundwater which is disinfected by ozone (not chlorine) as it is bottled, or is municipal tap water which has been treated with granular activated carbon (to remove chlorine and other dissolved organics), and with deionizing resins (to remove inorganics). Specific levels of minerals then are added, and the water is ozonized as it is bottled. As a result, U.S. bottled waters do not contain the chlorinous tastes and odors sometimes prevalent in U.S. municipal tap waters.

On the other hand, European tap waters also are either clean groundwaters, or are surface waters which are treated by a combination of processes such that they do not contain chlorinous tastes and odors. As a result, in Europe the bottled water with which we are familiar in the U.S. is not an alternative to municipal tap waters.

Europeans do drink "bottled water", but these are natural waters from mineral springs, which are marketed for their health benefits. Many terms are used for European bottled water, the most common being 'mineral water'. However, regulations recently adopted by member states of the European Economic Community now have standardized on the term, "natural mineral water".

In most European countries, about 80% of bottled mineral waters contain carbonation, either naturally or added. Carbonation is believed to add 'freshness' to the product, but it is also well established that common microorganisms cannot proliferate in carbonated water.

France is the major Western European producer of bottled water, but its product mix is reversed from those of the other countries. More than 80% of the French mineral water produced each year is 'flat', or still water, and only about 20% is carbonated. By contrast, about 80% of mineral waters bottled in Germany and Italy contain carbonation.

Each spring source of natural mineral water must be approved and is regulated by the appropriate national agency responsible for public health. Bottling normally is conducted at the source, which must be protected against "all risk of pollution". Disinfection by any means that is likely to change the viable colony count of approved natural mineral waters is prohibited. This is because counts of naturally present bacterial colonies are used as routine indicators of pollution. Without a colony count, much more sophisticated analytical techniques would be required to determine the presence or absence of pollution.

Many natural mineral waters have known health benefits which have been confirmed and approved by national academies of medicine and by public health organizations. These proven benefits to human health can be suggested in advertisements, but not claimed. Each label must carry the source and mineral analysis of the water.

In this paper, the authors describe the growth of Western European bottled water markets over the past decade, discuss the unique features of this market that distinguish it from the U.S. bottled water markets, then discuss the current regulations for bottled water in the European Economic Community countries, and compare these regulations with those currently extant in the U.S.A.

I. TYPES OF EUROPEAN BOTTLED WATER

In Europe, bottled drinking water generally is referred to as 'mineral water'. This is because its origin is subterranean groundwater, rather than being processed tap water. Groundwater contains naturally occurring minerals, hence the origin of the term.

Most European mineral waters are sold in 0.7-L bottles (glass or PVC), with the balance consisting of 0.25, 0.5, and 1.5 L containers. Essentially no mineral waters are sold in 5-gallon carboy type containers as is the common practice in the USA.

There are essentially three types of mineral waters in Europe, carbonated, non-carbonated, and 'health' or medicinal waters. Most of the bottled water consumed in Europe is carbonated (Germany 86%, Italy 80%). Much of the European population has been brought up with the understanding that carbonation imparts 'freshness' to the water. However, it is a well-known fact that common microbiological organisms cannot live in such waters because of the presence of carbon dioxide. Some non-carbonated mineral waters are deliberately carbonated to provide the desired degree of 'freshness'. Concentrations of carbon dioxide in these mineral waters is on the order of 8 to 10 g/L.

By contrast, the ratio in France is reversed. More than 80% of French bottled mineral waters are flat, or non-gaseous. In France, the only requirement to be called a "mineral water" is that it be from an underground source and have a proven therapeutic benefit.

So-called 'health' or medicinal waters contain traces of specific minerals, deliberately added (or removed), or occurring naturally, which provide some special health benefits. For example, mineral waters containing small quantities of fluoride are beneficial in combatting dental caries. Other minerals which impart special health benefits include iodine, iron, magnesium, calcium, and others.

Some medicinal waters contain high concentrations and others low concentrations of sodium chloride and sulfates. Others are carbonated, but to concentrations less than 2.5 g/L of carbon dioxide.

II. SOURCES OF EUROPEAN MINERAL WATERS

Each individual European country (and the EEC countries as a unit) requires that a natural mineral water must have its origin in a subterranean source. It must differ from tap water by its "original purity", special quality, mineral content, and its special effects. The mineral water must contain at least 1,000 mg of minerals per kilogram, or if not, it must prove its special effect by physiological, pharmacological, and clinical tests.

"Original purity" means, for example, that the waters contain no pesticides or halogenated organic compounds, and must be free of environmental contamination and pathogenic microbiological impurities.

France has about 1,500 known sources of naturally occurring mineral waters, of which about 900 currently are being used. Of these, only 25 are approved by the French Ministry of Health for bottling and sale as mineral waters which provide special therapeutic benefits to humans. Naturally occurring mineral waters from the other sources (which do not have demonstrated therapeutic properties) are bottled and sold as 'spring waters', at a much lower price than mineral waters with approved therapeutic properties.

Most of the French mineral water sources are in mountainous areas in the Vosges, Pyrenees, and Alps. Of the 1,500 French sources, about 100 are spas, where people come for recreation, but also to drink the therapeutic mineral waters directly from the springs.

In the Federal Republic of Germany, there are 500 mineral water sources which are owned by the 190 member companies of the VDM (Verband Deutscher Mineralbrünnen = Association of German Mineral Springs). These 190 companies supply most of the German mineral water market. The largest German companies produce more than 300 MM bottles/year, primarily 0.7-L in size. There are also 160 other German mineral water suppliers, each of which produces a maximum of 1 MM bottles/yr.

Italy has 200 mineral water producers which own 300 mineral water sources. The largest four of these control 30% of the Italian market. The top 20 Italian companies control 33% of the market; thus 67% of the Italian market is shared by the 180 smaller companies.

III. PRODUCTION OF EUROPEAN MINERAL WATERS

Western European mineral waters must be bottled at the source, except with special approval. Each spring has its own individual name, therefore one bottling company may offer more than one mineral water, since mineral contents and therapeutic benefits normally are unique to each individual spring.

Table I shows the number of major mineral water bottlers in seven European countries. The Federal Republic of Germany, Italy, Spain, and France have the highest number of mineral water bottlers.

Table II shows the annual production of mineral waters for the period 1971 through 1980 in France, the Federal Republic of Germany, Italy and Belgium (EEC Countries), Spain, Yugoslavia, Switzerland, Austria, and Portugal. Production in the four EEC countries has grown from 4,900 MM liters in 1971 to 7,297 MM liters in 1980, a growth rate of 49%. At the same time, production in the other five listed European countries has grown from 966 MM liters in 1971 to 1,898 MM liters in 1980, a 96% growth rate. In these nine listed European countries, total production grew from 5,866 MM liters in 1971 to 9,195 MM liters in 1980, an overall growth rate of 56%.

TABLE I. MAJOR MINERAL WATER PRODUCERS IN EUROPE

Country	vol. mineral water produced in 1980 MM Liters	no. significant producers
Austria	249	8
Belgium	427	44
Federal Republic of Germany	2,380	190 (350 total)
France	2,980	36
Italy	1,510	40 (200 total)
Spain	785	50
Switzerland	216	9

The percentages of mineral water production per country in 1980 are listed in Table III. For that year, France led with 32.4%, closely followed by the Federal Republic of Germany with 25.9%. The four listed EEC countries together produced 79.3% and the other five countries accounted for 20.6% of the nine country total production.

TABLE III. PERCENTAGES OF EUROPEAN MINERAL WATER PRODUCTION - 1980

Of the total 9,194.8 MM liters produced by nine countries:

EEC Countries	
France	32.4%
Federal Republic of Germany	25.9%
Italy	16.4%
Belgium	4.6%
Non-EEC Countries	
Spain	8.5%
Yugoslavia	5.4%
Austria	2.7%
Switzerland	2.3%
Portugal	1.7%

TABLE II. PRODUCTION OF MINERAL WATER IN WESTERN EUROPE, in millions of liters

country	1971	1972	1973	1974	1975	1976	1977	1978	1979	1980
France	2,554.0	2,720.0	3,097.0	2,955.0	2,815.0	2,950.0	2,827.0	2,945.0	3,076.0	2,980.0
Federal Rep. of Germany	912.0	960.0	1,150.0	1,197.0	1,496.0	1,870.0	1,947.0	2,095.0	2,275.0	2,380.0
Italy	1,200.0	1,200.0	1,344.0	1,183.0	1,089.0	1,197.0	1,260.0	1,350.0	1,560.0	1,510.0
Belgium	234.1	241.6	266.0	249.0	288.0	325.0	341.8	370.5	408.0	426.7
EEC TOTAL	4,900.1	5,121.6	5,857.0	5,584.0	5,688.0	6,342.0	6,376.0	6,760.5	7,319.0	7,296.7
Spain	225.0	278.6	336.7	351.5	315.0	417.0	384.0	422.0	740.0	785.0
Yugoslavia	460.9	503.7	485.0	490.0	436.4	425.1	444.9	460.2	492.3	495.6
Switzerland	154.6	157.4	176.0	174.9	187.6	205.5	196.4	195.9	207.0	215.8
Austria	56.2	59.2	87.0	102.0	126.0	167.5	192.4	212.0	248.0	248.5
Portugal	69.4	75.6	93.0	80.9	94.0	119.4	117.4	127.6	137.6	153.2
Total Europe	5,866.1	6,196.1	7,034.7	6,783.3	6,847.0	7,676.5	7,711.1	8,178.2	9,143.9	9,194.8

Medicinal waters in the Federal Republic of Germany currently have captured 10 to 15% of the total German mineral water market -- about 400 MM 0.7-L bottles (280 MM L). Two companies control 75% of this medicinal water market, according to the German Health Water Production Union, which consists of about 35 member companies.

IV. USES AND CONSUMPTION OF BOTTLED WATER IN WESTERN EUROPE

Each country has its own regulations concerning the uses for bottled water. For example, historically mineral water is allowed to be used for the production of soft drinks in the Federal Republic of Germany, but not in France.

Table IV lists the 1980 per capita consumption of mineral waters for the nine major European countries for which data are available. Consumption per capita in France was highest (nearly 56 L), followed by Belgium (43.5 L), then Germany, Switzerland and Austria in the 33 to 39 L/capita range, with consumption in Italy, Spain, and Yugoslavia being in the range of 21 to 27 L/capita. Consumption in Portugal was the lowest, 15.5 L/capita.

TABLE IV. CONSUMPTION PER CAPITA OF MINERAL WATER IN SOME EUROPEAN COUNTRIES

Country	Production in MM liters	Inhabitants (millions)	1980 Consumption in liters/capita
EEC Countries			
France	2,980	53.3	55.9
Fedl. Rep. of Germany	2,380	61.4	39
Italy	1,510	56.7	26.6
Belgium	427	9.8	43.5
Non-EEC Countries			
Spain	785	37.1	21.2
Yugoslavia	496	22.0	22.5
Switzerland	216	6.3	34.3
Austria	249	7.5	33.2
Portugal	153	9.9	15.5
Totals	9,196	264.0	35 (av)

It should be appreciated that in Europe bottled water is consumed directly for drinking, as opposed to being used to make coffee, tea, soups, soft drinks, general household cooking, etc. Normally tap waters are used for these purposes, especially in countries other than

France. This is because most of the bottled water is carbonated in these other countries.

It should be noted also that European tap waters do not have strong chlorinous tastes. Therefore, coffee, tea, soups, etc., prepared from European tap waters have tastes which are desired by European palates. These consumable liquids will taste the same in Europe whether prepared from flat bottled water or tap water.

V. BOTTLED WATERS IN THE MAJOR WESTERN EUROPEAN COUNTRIES

A. France

1. Introduction and Definitions

About 90% of the French mineral water market is served by four organizations which are composed of many of the 36 major French producers. Whereas most of the mineral waters produced in the remainder of Western Europe are carbonated (86% in Germany; 80% in Italy), the situation is reversed in France. During the 10 year period of 1971 through 1981, production of carbonated mineral waters in France ranged from 14.6 to 19.6% of the total mineral water produced.

In France, three types of drinking waters have been defined:

1) Potable Water, which, in turn, is comprised of three types:

   a) tap water, delivered to the public through piping systems

   b) spring water - naturally occuring from underground springs

   c) 'table water' - bottled water which is treated or untreated tap water. However, this type is disappearing, because French regulations do not currently require a statement noting that the water has or has not been treated.

2) Gasified Water, which contains added carbon dioxide,

3) Mineral Waters, which can be the same as spring waters, but with demonstrated therapeutic values.

Gasified waters (spring or tap waters to which carbon dioxide has been added) should not be confused with gaseous waters. The latter are naturally carboeffervescent.

Waters containing no carbonation (natural or added) are called flat waters.

Mineral waters are defined by the decree of 12 January 1922, as modified by the decree of 24 May 1957 as follows:

> "The terms 'mineral water' and 'natural mineral water' and all others containing these words are reserved for waters endowed with therapeutic properties, coming from a spring whose exploitation has been authorized by ministerial decision, under the conditions provided for by the laws and rules in force".

In contrast with legislation of numerous other Western European countries (Switzerland, Federal Republic of Germany), French authorities do not take into consideration the mineral content of the water to confer upon a spring the label 'mineral water'. Demonstrated and proven therapeutic properties (confirmed by the French Academy of Medicine) are all that is required.

> "By contrast, the term 'spring water', or all other (definitions) indicating a drinking water of determined origin, is reserved for potable water, that is, water suitable for human consumption, introduced (to the consumers or into commerce) at its place of emergence (from the ground)".

2. Sources And Current Market Characteristics

Currently, there are 1,500 sources of mineral waters in France, of which about 900 are active. However, only about 25 are approved by the French Ministry of Health, after notice from the French Academy of Medicine, as being beneficial to health.

The French mineral water industry grew rapidly during the period 1950 to 1970. In 1940, the total French production of mineral water was only 30 MM liters. By 1957, this had grown to 1 billion liters, to 2.5 billion liters in 1971, and to its peak of almost 3.1 billion liters in 1973.

During 1981, the total production of bottled mineral waters in France was 2,396 million liters. Of this amount, 2,000 MM liters (83.5%) were flat mineral waters, and 396 MM liters (16.5%) were gaseous. Over the ten year period 1971-1981, French production of bottled waters has remained relatively static, varying from a low of 2,054 MM L in 1971 (80% flat water/20% gaseous) to a peak of 3,097 MM L in 1973 (84% flat waters/16% gaseous).

The French bottled mineral water industry is served by 36 individual producers, but many of these are members of four major groups, which collectively control nearly 90% of the French market. These four groups, with their major product labels, are:

Perrier: Contrex, Perrier, Vichy Etat, St. Yorre, Célestins

B.S.N.: Evian, Badoit

Vittel: Vittel Grand Source, Hépar, Vitteloise, Pierval

Sellier Leblanc: Volvic, Volvillante

For 1981, French mineral waters were bottled in the following percentages:

Flat Waters

Perrier	Contrex	25.7%
B.S.N.	Evian	22.8%
Vittel	Vittel	17.1%
	Pierval	2.0
	Hépar	1.0
Volvic	Volvic	8.9%
Spring Waters	Various	17.8%
	Part-time distributors	4.7%
		100.0%

Mineral Waters, authorized by the Ministry of Health of the particular Department (State):

Gaseous Waters

Perrier	Perrier	39.3%
	St. Yorre	25.7%
	Vichy Célestins	5.5%
B.S.N.	Badoit	14.6%
Vittel	Vitteloise	4.0%
Various		10.9%
		100.0%

Contrexéville (Perrier), Vichy, Evian, and Vittel bottle nearly all French spring waters.

There has been little change in these relative positions since 1975, except that production of Volvic water has grown to its current level since that time.

3. Distribution

About 72% of French mineral waters are distributed through food stores, and 28% through cafes, hotels, and restaurants. In food distribution channels, about 37.8% is distributed through supermarkets, 27.5% in hypermarkets (many times the size of our supermarkets), 13.2% in superettes (small supermarkets) and 21.5% in other 'traditional' retail outlets other than cafes, hotels, and restaurants (such as pharmacies, department stores, very small neighborhood grocery stores, etc.).

4. Exports

Only relatively small quantities of French mineral waters are exported. For example, in 1981, 2,548 MM L were consumed domestically, with an additional 432 MM L (14%) exported. Of this total, 2,109 MM L were flat waters consumed domestically with an additional 290 MM L (12%) exported, and 439 MM L of gaseous mineral waters were consumed domestically, with an additional 141 MM L (24%) being exported.

5. Beverage Consumption in France

The average Frenchperson consumes annually:

100 L of wine

60 L of coffee

60 L of bottled water (53 L mineral water; 7 L spring water)

49 L of beer

Consumption of bottled 'table waters' is decreasing, because the high quality of the much lower cost French tap water is being increasingly recognized. For example, in late 1975, the Department of Health of the City of Paris advised all public assistance hospitals (then using bottled 'table water') that they should change immediately to tap water. The Health Department pointed out that Paris tap water 'is at least as bacterially pure as bottled table water, and it is considerably cheaper". This conversion had been "test marketed" and approved by both patients and doctors. The conversion from bottled water to tap water also allowed the hospitals to institute a regimen of changing the bedside carafes of water two times/day, rather than less frequently with the more costly bottled table water.

6. Therapeutic Uses of Mineral Waters in France

French mineral waters must have a therapeutic benefit in order to be approved by the Ministry of Health for sale as mineral water. Such therapeutic advantages must be certified by the French Academy of Medicine. Of course, the therapeutic value of specific mineral waters depends entirely upon the mineral contents of the waters. Once the therapeutic benefit of a particular mineral water has been certified and approved, the bottler is allowed to advertise this fact, which allows him to charge a higher sales price.

Several French mineral waters are listed below, along with their major mineral contents, their approved therapeutic benefits, and the approximate recent annual rate of sales:

Flat Mineral Waters

Contrexéville, contains high calcium sulfate (up to 1,500 mg/L), low contents of calcium bicarbonate, low sodium and low magnesium sulfate. About 500 MM L/yr are sold as a diuretic aid, also for treatment of biliary conditions and lithiasis.

Evian, is weakly mineralized, contains very low sodium (5 mg/L), 78 mg/L calcium, 24 mg/L magnesium, 357 mg/L bicarbonate ion. About 500 MM L/yr are sold for treatment of gout, arthritis, obesity, and renal infections.

Hépar, is rich in magnesium (110 mg/L), calcium (596 mg/L), and bicarbonate (403 mg/L). It is advertised for treating migraine headaches, hepatobiliary conditions, and excess cholesterol. Because of its high magnesium content, this mineral water is claimed to be a dietetic substitute for chocolates and dried fruits, which are high in magnesium, but also high in calories and are sometimes banished from human diets. This mineral water is claimed to replace magnesium required by the body to prevent fatigue and nervousness.

Vittel contains very low sodium (3 mg/L), 202 mg/L calcium, and 36 mg/L magnesium. About 554 MM L/yr are sold for treatment of gout, excess uric acid, obesity, urinary infections, and renal lithiasis.

Volvic has very low mineral content, only 10.4 mg/L calcium, 6 mg/L of magnesium, 8 mg/L of sodium, and 64 mg/L of bicarbonate. About 130 MM L/yr are sold as a diuretic, and for treatment of lithiasis, obesity, and improving metabolism.

Gaseous Mineral Waters

Badoit, contains 157 mg/L calcium, 1.3 mg/L fluoride, 83 mg/L magnesium, and 138 mg/L sodium. Its high sodium content makes it incompatible with low-sodium diets. About 75 MM L/yr are sold to improve digestion, and for preventing dental caries. It is a very agreeable water to drink, not too gaseous, which accompanies food very well when the diner wishes to avoid alcohol and drink other liquids.

Perrier, is only slightly mineralized, containing principally 140.2 mg/L of calcium. There are no major therapeutic benefits, other than its digestive qualities. It has very low sodium content (5 mg/L) and about 350 MM L/yr are sold.

Vichy-Etat is highly mineralized, containing variable amounts of calcium, magnesium, lithium, iron, and potassium, but very high sodium contents (1,280 mg/L). About 40 MM L/yr are sold for treatment of hepatobiliary disorders (after viral hepatitis, for example), desensitizing alimentary and digestive allergies, improving digestion, and protection of gastric fluids during treatment with antibiotics or cortisone.

Vichy-St. Yorre, also contains very high sodium contents (1,676.7 mg/L). About 200 MM L/yr are sold for the treatment of liver ailments caused by toxic substances, as a diuretic, for elimination of excess uric acid and cholesterol, and as a digestive aid. In spite of its high sodium content, it does not have a salty taste.

Spring Waters have essentially the same mineral contents as the above listed Mineral Waters, but do not have known therapeutic properties. As a result, they cannot be marketed as Mineral Waters. Therefore, their sale prices are much lower than prices commanded by mineral waters.

Table Water today is consumed mainly in the Northern sections of France, primarily where city water supplies are not yet available. However table waters may or may not be treated waters from any source - there is no current regulatory requirement that they must be treated.

7. Costs

In France, the consumer pays about 500 times the cost of tap water for bottled mineral water. The average cost of a 0.7-L bottle to the distributor is 1.50 French francs (Ff - currently the exchange rate is more than 8 Ff per U.S. dollar), broken down as follows: (total 1.50 Ff)

packaging cost (of filled bottles)	0.40 Ff
bottling	0.25
reimbursable Value Added Tax	0.25
non-reimbursable V.A.T.	0.10
general expenses	0.10
specific water rights	0.05
community taxes	0.02
profit	0.33

Transportation costs are in addition to the 1.50 Ff 'production' costs listed above. When bottled in glass, transportation costs are about 19% additional, and about 10% additional with plastic bottles. Both of these figures depend upon the distance of transportation from the source of the water.

About 54.6% of the cost of a bottle to the consumer is production cost when the bottles are plastic, and about 36.8% when the bottles are glass.

With mineral waters, advertisement of the specific brand name and its therapeutic benefits are the most important factors influencing sales. The therapeutic benefits of the water far outweigh the sales price in the eyes of the French consumer.

However with spring waters, which have no known or approved therapeutic benefits, price is the most important sales factor.

In light of this fact, a logical question is, "why does the spring water market share not increase?" The answer is that many springs have limited production rates, and only so much spring water can be bottled over a given period of time.

B. Federal Republic Of Germany

1. Introduction and Definitions

About 75% of the German population (14 years of age and older) regularly drinks mineral water. The market for mineral waters in Germany has increased rapidly, particularly from 1970 to 1981. Much of this recent growth has been attributed to an image change for mineral water, from that of a liquid to be drunk for its healthful properties only to a liquid which is refreshing to the taste and which quenches thirst. As recently as 1969, mineral water was a product purchased primarily by sportsmen, patients, and elderly people, primarily for health purposes.

Today, however, mineral waters have the image of Health plus Thirst/Refreshment. It is advertised as the kind of drink which is fresh (carbonation), healthy, without calories, and it is a drink which is required to be made available at parties and other activities where alcoholic beverages are served. Drivers of automobiles can lose their licenses permanently if they are caught driving 'under the influence', even one time. As a result, the car pool driver of party-goers usually limits his consumption to bottled water, fruit juices, or soft drinks.

About 86% of German bottled mineral waters are carbonated, either naturally or deliberately, with the balance being still waters. In contrast to France, however, German mineral waters are not limited only to those which have therapeutic uses. Those waters which do have documented therapeutic advantages are called _'health waters'_, or medicinal waters. In the past, mineral waters from a single spring have been sold both as a health water and as a mineral water (without health benefits). However, the new regulations adopted by the members of the European Economic Community countries in 1980 (see later discussion) forbid this practice.

Taffelwässern, or table waters are still or carbonated, but generally are simply waters bottled from any source, with the proviso that it is safe for human consumption.

2. Sources And Current Market Characteristics

The 1981 German market volume for bottled mineral waters was about 2,540 MM liters, which represents an average per capita consumption of approximately 42 L/year, consumed by the total population, 14 through 60 years of age. Consumption appears to be higher among more highly educated groups.

Although the Federal Republic of Germany has about 350 bottling companies and an unknown number of mineral water sources, the 190 member companies of the German Association of Mineral Water Producers (VDM = Verein Deutsche Mineralbrünnen) have 500 sources.

The major suppliers of mineral water (mostly in 0.7-L bottles) in the Federal Republic of Germany are as follows:

Apollinaris about 350 MM bottles (about 254 MM L)

Gerolsteiner Stern about 350 MM bottles (about 245 MM L)

Überkinger Group about 300 MM bottles (about 210 MM L)

(which consists of): Überkinger
Heigerloch/Bad Imnau
Bad Ditzenbach
Kissligg/Krumbach
Waiblingen/Beinstein

Blaue Quellen Group about 250 MM bottles (about 175 MM L)

(which consists of): Rhenser
Fürst Bismarck Quelle
Neustelters
Grauhof
Vorlo/Rietenau

Franken Mineral und Heilbrünnen Hufnagel KG/GmbH, Neustadt/Bavaria about 150 to 200 MM bottles (about 105 to 140 MM L)

An additional twelve companies each produce 50 to 130 MM bottles (about 35 to 91 MM L) per year, and 160 other medium and smaller companies each produce up to 1 million bottles/year (about 0.7 MM L).

The top ten German bottled mineral water companies have about 25% of the current market, and the top 25 companies have a market share of about 45%.

3. Distribution

In the Federal Republic of Germany, 65% of mineral waters produced are sold to households, 20% to restaurants, and 15% to cafeterias of industrial companies and other sources. About 66% is distributed by wholesale dealers of beverages (who also deliver to food dealers), about 30% by discount purchases, and about 4% by home service, delivered by small trucks.

Bottle sizes and percentages of their use are as follows:

0.7 L approx. 76%

0.25 L approx. 15% (esp. for restaurants)

0.5 L approx. 7%

1.0 L & others approx. 2%

Whereas the number of aluminum cans used for soft drink packaging is growing in Germany, the use of cans for mineral water is less than 0.001% of that for bottles at present.

4. Exports and Imports

About 60% of the total German mineral water exported (about 200 MM L/yr) is bottled by the largest producer, Apollinaris, which exports on a world-wide basis. Most other German exporters sell primarily to other European EEC countries.

In spite of recent aggressive marketing efforts of French producers, only some 3% (about 78 MM L) of mineral waters consumed in the Federal Republic of Germany are imported, at the present time.

5. Consumption of Beverages in the Federal Republic of Germany

During 1981, the average German consumed some 661.7 liters of beverages, divided as follows:

coffee and substitutes	206.5 L
beer	147.0
milk	93.4
soft drinks	69.0
tea	49.8
mineral water	41.9
wine	20.2
spirits	7.7
champagne	4.4
total	661.7 L

Although the 1981 consumption rate of mineral water in Germany was about 42 L/person, it has been projected that consumption can grow to a maximum of 55 L/person/year by 1990. The figure of 55 L (14.5 U.S. gallons)/person/year currently is believed to represent a maximum saturation of the potential German market. It is based on the fact that French consumption of mineral waters since 1973 has remained more or less fixed at about this rate.

At first glance, the per capita consumption of 55 L/year appears to be a very low maximum water consumption figure. However, it should be pointed out that mineral waters are consumed directly. It is estimated that the average German consumes an additional 2.4 L/day of tap water used for making coffee, tea, and soups. This equates to an additional 912.5 L/yr, or a total of 967.5 L/yr, which converts to 255 U.S. gallons per year.

The senior author of this paper has been purchasing non-carbonated bottled water since 1979. For his family of two people, he purchases an average of 25 gallons per month (300 U.S. gallons/yr), in five-gallon plastic bottles. However, bottled water is used not only for drinking, but also for cooking, reconstitution of powdered milk, making coffee, tea, lemonade, etc. In Germany (and other European countries except France), because most mineral water is carbonated, it does not lend itself to cooking, or the other purposes mentioned.

Table V shows the development of the German mineral water market from 1938 through 1981. The increase in consumption from 2 L/person/year in 1938 to about 42 L in 1981 is particularly striking.

TABLE V. DEVELOPMENT OF GERMAN MINERAL WATER MARKET, 1938-1981

year	consumption/capita/yr in liters	market volume in Germany
1938	= 2.0 L	140 MM L
1950	= ca. 3.8 L	190 MM L
1970	= ca. 13.0 L	766 MM L
1981	= ca. 42.0 L	2,540 MM L

In 1970 the share of mineral water consumption in Germany was 2.4% of the total beverage market (517.6 L/capita - of which beer represented 141 L/capita). In 1981, the share of mineral water had risen to 6.3% of the total German beverage market (662 L/capita - of which beer represented 147 L/capita). Consumption of mineral water more than doubled during this time, while per capita beer consumption rose by slightly more than 4%. Still, however, Germans drink more than three times the volume of beer than bottled water.

6. Medicinal Waters in the Federal Republic of Germany

About 400 MM 0.7-L bottles of health waters are produced each year at the present time. This represents about 10 to 15% of the current German mineral water market. About 45% of this production is made by the Uberkingen group, and 30% by the Fachingen company (about 120 MM bottles).

There is a German Association for Health Water, which has some 35 members. These waters are distributed by wholesale dealers for food and discount stores, and are sold in pharmacies, drugstores, and hospitals.

7. Costs of Mineral Waters in the Federal Republic of Germany

The average price paid for an individual 0.7-L bottle of mineral water in 1973 was 0.54 DM; 0.49 DM in 1977; 0.50 DM in 1979; and 0.52 DM in 1981/1982. During late 1982, the price charged for a box of 12 0.7-L bottles was 6.60 DM (0.55 DM/bottle, including box deposit). The current conversion rate is about 2.7 DM per U.S. dollar.

Prices charged in German stores for a box of 12 0.7-L bottles of three classes of mineral water (inexpensive, middle-class and high priced) during late 1982 were as follows:

Inexpensive Class	1.99 - 3.00 DM/box	(0.16 - 0.25 DM/bottle)
Middle Class	3.50 - 5.00 DM/box	(0.29 - 0.42 DM/bottle)
High Class (Apollinaris, Gerolstein, etc.)	5.00 - 9.00 DM/box	(0.42 - 0.75 DM/bottle)

Medicinal water costs in Germany range from 0.65 to 1.55 DM/0.7-L bottle (high priced brand = Fachingen).

Apollinaris' high priced brand sold in 1-L bottles costs 0.95 to 1.20 DM/bottle.

Costs for 1.5-L PVC bottles of various imported mineral waters in Germany are as follows:

Spa	(Belgium)	=	0.80 - 0.90	DM/bottle
Vittel	(France)	=	0.85 - 0.95	do.
Contrex	(France)	=	1.15 - 1.20	do.
Evian	(France)	=	0.85 - 0.95	do.
Highland Springs (Great Britain)		=	about 1.00	do.

Production Costs of Mineral Water in the FRG

Costs in DM/bottle for production, bottling, sales and distribution of mineral water in the Federal Republic of Germany have been estimated to be as follows:

labor costs	=	0.08 - 0.10 DM/bottle
freight costs		
200 km	=	0.09
400 km	=	0.14
800 km	=	0.21
advertising and sales	=	0.03 - 0.06
costs for bottle, label, etc.	=	0.05 - 0.15
		0.16 - 0.31 DM/bottle, plus freight costs

One benefit accruing to the 190 member companies of the VDM is that the 0.7-L glass bottles are standardized and interchangeable. This means that each member company is able to use the 0.7-L bottles of any other member company.

C. Italy

In this country, there are about 200 mineral water bottlers, who control about 300 sources of mineral waters. However, most of these are small companies. The top four companies control only 30% of the Italian mineral water market, 33% of this market is controlled by the largest 20 Italian bottlers, and 67% of the market is served by the smaller 180 companies.

Carbonated mineral waters account for about 80% of the Italian market; however, only 0.3% of the natural mineral waters contain carbon dioxide at the source. Therefore, most Italian mineral waters must be carbonated. The two companies who sell the naturally carbonated mineral water advertise it as being 'naturally carbonated'. However, Italians who drink carbonated water want more carbonation than is present in the naturally carbonated water. Thus, even the naturally carbonated water usually is supplemented with carbon dioxide.

The four major Italian mineral water producers and their individual shares of the recent mineral water market are as follows:

St. Gemini	10%
St. Pellegrino	10%
Bognanco	5%
Recoraro	5%

St. Gemini's mineral water has been approved for preparing babies milk formulas.

Consumption of mineral waters by Italians is about 1 L/person/week for about 50% of the Italian population. The other 50% do not consume bottled mineral water. Most of the Italian mineral water is sold in glass bottles, with maximum capacity of 1-liter. There is one plastic bottle, of 2-L capacity. There are no 5-gallon bottles. Instead, the consumer buys a box of 12 1-L bottles.

Very little Italian mineral water is exported.

VI. REGULATION OF WESTERN EUROPEAN BOTTLED WATERS

A. The EEC Uniform Standards For Natural Mineral Waters

On 30 August 1980, the Official Journal of the European Communities published two Council Directives, both of which had been adopted by the member countries on 15 July 1980. These Directives relate to:

(1) The Quality of Water Intended For Human Consumption (Official Journal No. L. 229/1), and

(2) The Approximation of the Laws of the Member States Relating to the Exploitation and Marketing of Natural Mineral Waters (Official Journal No. L 229/11).

The Directive relating to the Quality of Water Intended For Human Consumption does not apply to "natural mineral waters recognized or defined as such by the competent national authorities, nor to medicinal waters recognized as such by the competent national authorities". This directive is concerned with agreement upon parameters of potable water to be measured, frequency of measurement of these parameters, and reference methods of analysis for these parameters. In addition, Guide Levels and Maximum Allowable Concentrations for each parameter also are listed, where agreement has been reached.

The Directive relating to the Exploitation and Marketing of 'Natural Mineral Waters', defines that term, and others, then lists requirements and criteria for applying the definitions, supplemental requirements for naturally effervescent natural mineral waters, and conditions for the exploitation and marketing of natural mineral waters. Of most importance is the agreement reached for definition of terms, particularly use of the term, "natural mineral water", and the conditions under which a source for such water can be commercialized, bottled, labelled, analyzed, and regulated.

There are thirteen important features of the regulations adopted by the EEC for bottled water:

1) <u>Uniformity Of Definition</u>

"<u>Natural Mineral Water</u> means microbiologically wholesome water (within the meaning of Article 5 of the Directive - see item 6 below, Microbiological Regulations), originating in an underground water table or deposit and emerging from a spring tapped at one or more natural or bore deposits".

The term "mineral water" no longer can be used.

2) <u>Sources of Natural Mineral Waters</u>

As indicated by the above definition, natural mineral water originates in an underground source. It is distinguished from ordinary drinking water by its nature (characterized by mineral content and, where appropriate, by its medicinal effects), and by its original state.

Both of these characteristics are preserved intact because of the underground origin of the water, and because the source is required to be "protected from all risk of pollution".

The spring source must be approved by the responsible health authority in the Member State. Certification of approval must be obtained from the responsible health authority every two years.

In approving a source of natural mineral water, the responsible health authority shall conduct a complete survey of the source, which includes determining the rate of flow of the spring, temperature of the water at source and the ambient temperature, relationship between the nature of the terrain and nature and types of minerals in the water, dry residues at 180°C and 260°C, electrical conductivity or resistivity (with the temperature of measurement having to be specified), pH, anions and cations, non-ionized elements, trace elements, radiological properties at source, the relative isotope levels of the constituent parts of water (isotopes of oxygen and hydrogen, where appropriate), and the toxicity of certain constituent elements of the water, taking into account the limits set for them.

The catchment, pipes, and reservoirs must be of materials suitable for water, and so built as to prevent any chemical, physico-chemical, or microbiological alteration of the water.

3) Listing of Natural Mineral Waters

The Directive calls for a listing of waters extracted from the ground in Member States and recognized by the responsible authorities of those Member States as natural mineral waters. In addition, those natural mineral waters imported from a third country (outside of the EEC) and recognized as natural mineral waters by the responsible authority of a Member State also are required to be listed in the Official Journal.

Through mid-July, 1983, however, no such listing had appeared.

4) Allowable Treatments

No chemical treatment of natural mineral waters is allowed except for the following:

a) separation of its "unstable elements", such as iron or sulfur compounds, by filtration or decantation, possibly preceeded by oxygenation, as long as this treatment does not alter the composition, as regards the essential constituents of the water which give it its properties;

b) total or partial elimination of free carbon dioxide by exclusively physical means;

c) introduction or reintroduction of carbon dioxide under conditions specified on the label. The waters may be termed "naturally carbonated natural mineral water", "natural mineral water fortified with gas from the spring", or "carbonated natural mineral water".

5) Prohibition of Disinfection

Disinfection treatment, by whatever means, and the addition of bacteriostatic elements, or any other treatment which is likely to change the viable colony count of the natural mineral water is prohibited. The reason for this is that the natural viable colony count is considered to be a measureable indicator of pollution. If the count is lower than that which normally occurs at the source, this can be interpreted to be an indication that the source has become polluted. If the count is significantly higher than at the source, this indicates that the water could have become polluted, and/or could have been subjected to elevated temperatures during storage or transit.

The qualitative and quantitative composition of the normal viable colony count of the natural mineral water must be checked by periodic analysis.

6) Microbiological Regulations

Article 5 of the Directive requires that:

(a) After bottling, the total colony count at source may not exceed 100/mL at 20 to 22°C in 72 hrs (on agar-agar or agar-gelatine), and 20 per mL at 37° in 24 hrs (on agar-agar). The total colony count shall be measured within the 12 hours following bottling, with the water being maintained at 4°C ± 1°C during this time.

(b) At source, these values normally should not exceed 20/mL at 20 to 22°C in 72 hrs, and 5/mL at 37°C in 24 hrs, with the understanding that these figures are to be considered as guide numbers and not as maximum permitted concentrations.

(c) In addition, at source and during marketing, a natural mineral water shall be free of the following:

i) parasites and pathogenic organisms;

ii) *Escherichia coli* and other coliforms, and fecal streptococci in any 250 mL sample examined;

iii) sporulated sulfite-reducing anaerobes in any 50 mL sample examined;

iv) *Pseudomonas aeruginosa* in any 250 mL sample examined.

At the marketing stage:

i) the revivable colony count may only be that resulting from the normal increase in bacteria content which the natural mineral water had at the source;

ii) the natural mineral water may not contain any organoleptic defects.

7) Bottling

The washing and bottling plant must meet hygiene requirements. Bottles must be washed, disinfected, and final rinsed with water to be bottled.

8) Packaging

Containers must be treated or manufactured so as to avoid adverse effects on the microbiological and chemical characteristics of the natural mineral waters. Containers must be fitted with closures designed to avoid "any possibility of adulteration or contamination".

9) Advertising and Labelling

Sales literature and labels shall use the specifically approved terms, as well as any indications as to treatments (approved) which the waters may have undergone. Labels also must include the following mandatory information:

(a) Either the words: "composition in accordance with the results of the officially recognized analysis of ...... (date of analysis)", or

a statement of the analytical composition, giving its characteristic constituents;

(b) the location of the spring and its name.

It is forbidden to market natural mineral waters from one and the same spring under more than one trade description. It is also forbidden for a company to suggest a characteristic that the water does not possess. The term 'mineral water' no longer can be used.

Claims that a natural mineral water is effective for the prevention, treatment, or cure of a human illness no longer can be made. However, indications as to mineral content can be made as listed in Table VI, provided they have been drawn up on the basis of physicochemical analyses, and where necessary, pharmacological, physiological, and clinical examinations carried out in accordance with recognized scientific methods and in accordance with assessment of those characteristics which give natural mineral waters properties favorable to health.

These analyses should be suited to the particular characteristics of the natural mineral water and its effects on the human organism, such as diuresis, gastric and intestinal functions, and compensation for mineral deficiencies. These analyses are optional when the water already had been approved as (or had been considered to be a) natural mineral water in the Member State prior to entry into

TABLE VI. ALLOWABLE STATEMENTS FOR NATURAL MINERAL WATERS

Statement	Criteria
Low mineral content	Mineral salt content, calculated as a fixed residue, not greater than 500 mg/L
Very low mineral content	Mineral salt content, calculated as a fixed residue, not greater than 50 mg/L
Rich in mineral salts	Mineral salt content, calculated as a fixed residue, greater than 1,500 mg/L
Contains bicarbonate	Bicarbonate content greater than 600 mg/L
Contains sulfate	Sulfate content greater than 200 mg/L
Contains chloride	Chloride content greater than 200 mg/L
Contains calcium	Ca content greater than 150 mg/L
Contains magnesium	Mg content greater than 50 mg/L
Contains fluoride	Fluoride content greater than 1 mg/L
Contains iron	Fe(II) content greater than 1 mg/L
Acidic	Free CO_2 content greater than 250 mg/L
Contains sodium	Na content greater than 200 mg/L
Suitable for the preparation of infant food	----
Suitable for a low-sodium diet	Na content less than 20 mg/L
May be laxative	----
May be diuretic	----

force of this Directive. This is the case in particular when the water contains, both at source and after bottling, a minimum of 1,000 mg of total solids in solution, per kg, <u>or</u> a minimum of 250 mg of free carbon dioxide.

Member States may adopt special provisions regarding information - both on packaging or labelling and in advertising, concerning the suitability of a natural mineral water for the feeding of infants.

10) <u>Production of Natural Mineral Water</u>

Bottling generally is conducted at the source. Equipment must be installed so as to avoid any possibility of contamination and to preserve the properties which the water possesses at source. In particular:

(a) the spring or outlet must be protected against the risks of pollution;

(b) the catchment, pipes, and reservoirs must be made of materials suitable for water, and so built as to prevent any chemical, physicochemical or microbiological alteration of the water;

(c) the washing and bottling plant must meet hygiene requirements.

(d) transportation is permitted only in authorized containers, except if a Member State has authorized the transport of natural mineral water in tanks from the spring to the bottling plant at the time of notification of the Directive.

11) <u>In the Event of Pollution</u>

If the natural mineral water no longer presents the microbiological characteristics presented in Article 5 (see item 6, Microbiological Regulations, above), all operations must be suspended, particularly the bottling process, until the cause of pollution has been eradicated and the water again complies with the provisions of Article 5 (Microbiological Regulations).

12) <u>Monitoring</u>

The responsible authority in the country of origin shall carry out periodic checks to determine that the authorized conditions exist and are being applied.

13) <u>Export</u>

Requirements of the Directive do not apply to natural mineral waters intended for export to countries outside of the EEC.

VII. COMPARISON OF EUROPEAN AND AMERICAN BOTTLED WATER REGULATIONS

A. Similarities

With a few noteworthy exceptions, regulations applicable to the European bottled water industry are quite similar to those in force in the USA. Sources of water are to be approved by the appropriate public health authorities and protected from pollution. European source groundwaters are sampled and analyzed at regular intervals (every two months in France), to be sure that the mineral and bacterial contents have not changed. Pathogenic organisms must be absent in the product waters. European and American catchments, pipes and reservoirs must be made of materials suitable for the handling of water. Additional European requirements are that the catchments, pipes, and reservoirs be constructed so as to prevent chemical, physical-chemical, or microbiological alteration of water quality.

Bottling plants and facilities in both regions must be protected against entry of domestic and wild animals, rodents, and insects, and must be maintained hygienically clean. Water containers must be manufactured or treated to avoid adverse effects on microbiological and chemical characteristics. Containers must be rigorously inspected when they are received (new or returned) for contamination, particularly for their having been used as containers for chemicals. They must be washed, sterilized, and rinsed in product water before being filled.

Closures and capping equipment in both regions must be designed to prevent any possibility of contamination, and also must be washed, sterilized, and rinsed with product water before being used.

Product waters in both regions must be free of organoleptic defects.

B. Differences

The primary differences in European and American regulations are in microbiological characteristics of the waters, allowable treatments of the waters, and labelling requirements.

European bottlers must not destroy the natural, viable colony counts in their source waters. Since pathogenic organisms are absent, the viable total colony count of the natural waters can be used as an indicator of the presence of pollution. If the colony count falls, this is taken as an indication that some polluting chemical has appeared in the source which is toxic to the microorganisms present. Similarly, if the colony count suddenly rises, this can indicate the presence of a pollutant which serves as a bacterial stimulant. If higher than normal colony counts are found in the product water at any point in the distribution chain, this can indicate the same type of pollution, but also that the bottled water may have been stored under excessively high temperature conditions.

Europeans allow no chemical treatment of their waters, other than the possibility of oxygenation to assist in removal of unstable elements, such as iron or sulfur by filtration or decantation. Carbon dioxide may be added, but always by purely physical means. This restriction is necessary in order that the viable colony counts not be affected by chemical treatment.

American source waters can be underground or tap waters. If underground sources having high mineral contents are employed, they normally will be deionized in the USA, either by means of reverse osmosis or deionizing resins (or both). City tap waters will be dechlorinated by passage through granular activated carbon. In some cases, source waters will be distilled.

All American bottled waters are disinfected as they are being bottled, and this normally is accomplished by applying ozone as the terminal step in whatever treatment process has been adopted for the particular water.

European labelling requirements are more stringent than in the USA. This is because the properties of a specific natural mineral water are dependent upon the types and concentrations of specific minerals present in the source water. As a consequence, European labels must carry the name of the spring and its location, as well as a minerals analysis, or the statement, "in accordance with results of the officially recognized analysis of _______ (date)".

Prior to the adoption of the EEC regulations on 15 July 1980, each bottler whose source water had demonstrated therapeutic benefits for humans could claim those benefits, both in his labelling and in his advertising. Subsequent to adoption of the EEC Directive, however, such medical claims no longer can be printed on the labels. This also avoids the former practice, apparently widespread with some bottlers, of selling water from the same spring in two different bottles, one carrying the therapeutic advantages label and commanding a high price, and the other carrying a spring water label, and commanding a lower price.

In the USA, labelling requirements are specified by the Food & Drug Administration and by individual states. Some states adopt the FDA regulations, but others sometimes specify more stringent requirements. In general, American bottled waters do not have to carry a complete mineral analysis on the label, because most American waters have been demineralized.

However, in cases which involve the content of sodium, bottled American waters can list the sodium content to the nearest 5 mg/L. In fluoridated US bottled waters, the label should contain directions for use, rather than simply the fluoride content.

Monitoring is conducted periodically and routinely by the appropriate public health authorities in both countries, both national and regional.

One item of particular interest should be mentioned at this point. In Italy, mineral waters are considered to be natural resources. As a result, Italian bottlers of mineral waters are regulated by the Ministry of Industry, rather than by the Ministry of Public Health.

C. Regulation of Specific Chemicals in Bottled Water

In the United States, bottled waters are regulated by the FDA, which requires that bottled waters meet EPA's Primary Drinking Water Standards, or the Commissioner of FDA must publish the reasons for not requiring this. Table VII lists specific cations and anions and their maximum permitted levels in bottled waters of the USA, Federal Republic of Germany, and of France. Absence of a limit should not necessarily be interpreted to mean that there is no standard for such entries, but rather that such data have not been available to the authors.

In the Federal Republic of Germany, as in the USA and France, the drinking water standard for manganese is 0.05 mg/L. However, higher levels of manganese are permitted in German bottled waters, if such higher levels are present naturally in the source water. On this same basis of being present naturally, sulfate, chloride, and copper concentrations can exceed the drinking water standards in German bottled waters.

Table VIII lists the regulated organic chemicals in the USA and in the Federal Republic of Germany. Six pesticidal organics are regulated in the USA, and each has its own maximum limitation. On the other hand, Germany has a requirement limiting the concentration of _any_ single pesticide to 0.0001 mg/L each. The sum of all pesticides present must not exceed 0.0005 mg/L. In addition, polycyclic aromatics (total) are limited in Germany to 0.0002 mg/L, and six specific compounds are listed.

Finally, the US limits trihalomethanes to 0.10 mg/L, whereas in Germany, haloforms must be maintained at levels "as low as possible". One _microgram_ per liter is the current guideline for haloforms in Germany.

TABLE VII. MAXIMUM CONTAMINANT LEVELS OF BOTTLED WATERS

Contaminant	USA	Germany	France
As	0.05 mg/L	0.05 mg/L	0.05 mg/L
Cd	0.01	0.005	
Cl^-	250	200*	
Cr	0.05	0.05	not detectable
Cu	1.0	3.0**	1.0
Fe	0.3	0.2	0.1
Pb	0.05	0.05	0.1
Mn	0.05	0.05**	0.05
Hg	0.002	0.001	
Nitrate (as N)	10 (as N)	50 (as NO_3^-)	10 (as N)
Phenols	0.001	0.0005, not reacted with chlorine	must be absent
Se	0.01	0.010	0.05
Ag	0.05	0.010	
Sulfate	250	250*	
TDS	500	--	--
Zn	5.0	5.0	5.0
Ni	---	0.05	
Sb	---	0.01	
Cyanide	---	0.05	not detectable

* Higher levels are permitted if item is a natural component of mineral water.

** Drinking water standard. Higher levels are permitted for mineral waters.

TABLE VIII. REGULATION OF ORGANICS IN BOTTLED WATER

USA	Germany
Endrin 0.0002 mg/L	Pesticides, 0.0001 mg/L each
Lindane 0.004 mg/L	Pesticides, total, 0.0005 mg/L
Methoxychlor 0.1 mg/L	
Toxaphene 0.005 mg/L	
2,4-D 0.1 mg/L	
2,4,5-TP Silvex 0.01 mg/L	Polycyclic aromatics, total = 0.0002 mg/L: fluoranthene; benzo-3,4-fluoranthene; benzo-11,12-fluoranthene; benzo-3,4-pyrene; benzo-1,12-perylene; indene (1,2,3-cd) pyrene
Trihalomethanes 0.10 mg/L	Haloforms - as low as possible (1 microg/L is guideline)

VIII. CONCLUSIONS

European bottled waters are naturally occurring mineral waters obtained from groundwater sources and which contain significant mineral contents, which differ in identity and quantity from source to source. Some of these mineral waters provide specific and proven therapeutic advantages to humans. Because of this, European bottlers are forbidden to treat natural mineral waters to change their compositions. They are also forbidden to disinfect natural mineral waters, because the viable colony count at the source is used as an indicator of lack of pollution of the source waters.

Bottling is conducted as close to the source as practical, consistent with preventing the bottling plant from polluting the source. All materials used in bottling European natural mineral waters must be appropriate for handling water, the bottling plants must be maintained in hygienically safe condition, bottles and closures must be cleansed, disinfected, and rinsed with waters to be bottled, and filled and capped bottles must be maintained in hygienically safe conditions. Labels must name the source, its location, and provide a minerals analysis.

The cognizant public health authority of each European country has primary regulatory and monitoring responsibility for bottled water plants, but regional and local authorities sometimes have additional regulatory requirements, and perform routine monitoring.

In contrast to American practice, almost 80% of bottled European waters are carbonated (20% in France), and are sold in 0.7-L bottles, rather than in 5-gallon carboys.

IX. ACKNOWLEDGEMENTS

The authors are grateful to many European individuals and organizations who provided information necessary for the construction of this paper. In France, M. Pierre Schulhof, Director, Compagnie Générale des Eaux, Paris; in Italy, Dr. Giancarlo Riva, Technical Director, Ozono Elettronica Internazionale, Stradella; in the Federal Republic of Germany, Dr. Wilhelm K.G. Schneider, Director, and Dr. H. Kussmaul, Institut Fresenius, Taunusstein; Dr. W. Kühn, Engler-Bunte Institut der Universität Karlsruhe; Dipl. Ing. E. Wedde, Mannesmann Demag Hüttentechnik, Duisburg; and Dr. Eihoff, Bundesverband der Deutschen Erfrischungsgetränke-Industrie e.V, Bonn.

We are also grateful to William F. Deal of the International Bottled Water Association, and to Timothy L. Harker of Kadison, Pfaelzer, Woodward, Quinn, & Rossi, for suggesting this presentation, and for their guidance and counsel during its development.

Special thanks are due Mrs. Hanne Königsberg of Silver Spring, MD for translation of the German bottled water regulations.

X. BIBLIOGRAPHY

ANONYMOUS, "Resolution of 10 August 1961 (France), Modified by Resolutions of 28 Feb. 1962, 7 Sept. 1967, and 22 May 1973". Section I - General Regulations; Section III - Bottled or Conditioned Waters; Title II - Bottled Waters; Title V - Particular Regulations for Bottled Water. Direction des Journaux Officels.

ANONYMOUS, "Why Drink Water in Bottles?", 50 Million Consumers 44(8):8-15 (1974).

ANONYMOUS, "The Bottled Water Industry" (France), l'Eau Pure 46(Aug.-Sept.-Oct.):25-28 (1977).

ANONYMOUS, "New Regulations For Table Water", The Soft Drink and Mineral Water News (Federal Republic of Germany), undated, but late 1982, p. 382-392.

BRUN, P., "Bottled Mineral Waters - Their Therapeutic and Health Benefits", Bull. de l'Assoc. Pharmaceutique Français pour l'Hydrologie, July (1971).

CATS, M., The Drinking Water Industry: Regulatory, Economic and Social Aspects", Centre Belge d'Etude et de Documentation des Eaux 384(11):401-402 (1975).

CONSOLE, O., "Technical Training Course", Intl. Bottled Water Assoc., Alexandria, VA (1980).

CONSOLE, O., "Plant Technical Manual I", Intl. Bottled Water Assoc., Alexandria, VA (1982).

MINISTERE DE l'ECONOMIE, Annual Statistics of France, 1980, Vol. 85, Results of 1979, New Series No. 27, p. 244.

NINARD, B., "Conditioned Drinking Waters", l'Eau Pure 15(4):7-14 (1971).

OFFICIAL JOURNAL of the European Communities, "Council Directive of 15 July 1980 on the Approximation of the Laws of the Member States Relating to the Exploitation and Marketing of Natural Mineral Waters", Vol. L 229/1, p. 1-10, 30 August 1980.

PALLEZ, G., "Notice for the Directors of Hospitals and Groups of Hospitals", Public Assistance Hospitals of Paris, France, Directive dated 2 December 1975.

SIMPERE, F. "At the Seat of Experiment: Waters Full of Resources", Reference unknown, but published about 1982.

SCHNEIDER, W.K.G., "Market Study for Mineral Water", Inst. Fresenius, Taunusstein, Federal Republic of Germany, 1982.

THOMAS, A., "Bottled Waters: A Market Without Great Commotion", Strategies (France) 327:32-37 (16 Feb. 1982).

TRICARD, D., "Waters Destined For Human Consumption. Sanitary, Legislative and Regulatory Aspects", Travaux, Jan. 1979, p. 45-46.

CHAPTER 19

CONGRESSIONAL INITIATIVES

Honorable Robert T. Stafford

United States Senate
Washington, DC 20510

Thank you for permitting me to participate in your symposium on safe drinking water, an issue critical to the very survival of our nation. Like the seasons, environmental issues wax and wane. Slowly -- almost imperceptibly-- we move from one to another. And, also like the seasons, each environmental issue seems to return on schedule with a vigor and a force that had dimmed in our memory. As we seek to respond to these issues, the facts and the laws each become more and more complex.

The Clean Water Act began in the late 1940s as a federal process of negotiating water disputes. The Clean Air Act began in much the same way in 1954. Today those laws are complex regulatory mechanisms, and they become more complex each time we deal with them. The Clean Air Act, for example, is 190 pages long. And, if the reauthorization bill recommended last year by the Senate Committee on Environment were to become law, another 114 pages would be added to that total.

The reason for all this is, of course, that environmental issues rarely are as simple as they seem. When we enacted the various amendments to the Clean Water Act during the Decade of the Seventies it seemed a simple task to restore the quality of our rivers and streams. In some respects, it actually was. Great national waters -- like the Potomac River and Lake Erie -- have made remarkable recoveries, thanks in part to the massive expenditures of federal money. We have not been so successful with some other rivers and lakes, as you know. And, of course, it will be a long and costly struggle to achieve our goal of clean water. But, by and large, the remaining pure rivers of America are those hidden from sight. They are deep under the surface, and they run cleaner than the freshest snowmelt or glacial runoff. Sad to admit, now they, too, are threatened.

It is a genuine tragedy that America has exhausted so many of the mountain rivers of my youth. We have polluted and poisoned our surface waters to the extent that if I were to offer any one of you two glasses of water -- one drawn from a river and one drawn from the groundwater underneath-- you would make your decision without hesitation. But, it is an even greater tragedy that each of you would probably pause for an instant -- and, perhaps even longer -- to silently ask whether even the second glass of water -- the underground water -- was safe to drink. Experts that you are, you would reflect on the burden of the chemical and biological poisons dumped on the surface and wonder whether they had finally reached the well water -- as you know they inevitably must.

The first loss -- that of the mountain brooks and streams of my childhood -- is terrible. But, perhaps it is a loss that can be made up. It we stick to it, eventually -- and at a great cost -- we may turn back the clock. Children may again drift safely in a small canoe, as I did, or fish for the Atlantic salmon, as did my father and grandfather. Great sums of money, aided by the purifying and healing powers of the sun and with the wind, may bring those memories back to life.

But recovery may not be possible if we permit the great rivers of the underground to be poisoned. The sun does not shine on them, nor does the wind ripple their surface. They would be lost, quite possibly, forever. It is a loss we cannot afford. Such a loss would mean the squandering of one of our nation's greatest resources. It would be the spending of the wealth of our children because of our shortsightedness.

If you dig deep enough in the United States, you are likely to strike water. It is almost everywhere. Fully one-third of the nation lies over aquifers capable of yielding at least 100,000 gallons a day to a single well. Less productive aquifers underlie still more land. The supply of usable fresh water stored within the first half-mile of the surface of this nation is at least twenty times greater than the amount of water held in all rivers, lakes and streams, in the United States. Most of this water is still virtually pristine in quality. The total amount is even higher if we count all the water that is naturally salty, brackish, mineralized, alkaline, bad-tasting or otherwise unfit to drink. Still more lies at levels too deep to recover at today's prices.

It is a simple fact that the United States is groundwater rich. We are supplied by the world's largest solar-powered engine. Each year, precipitation puts back about ten times as much water -- 300 trillion gallons -- as we pump out of the ground. Although most rainwater either runs off into waterways or evaporates, as much as 30 percent of it seeps into the upper layers of the soil, then percolates down to the water tables. There the water runs through aquifers that are successively deeper and deeper. And there the water waits for man.

A two-inch pipe can be driven through the earth to supply a family with daily water needs. Or, in the arid West, a giant underground lake may be tapped to run a huge engine of irrigation turning dry and dusty earth into productive farmland. But, in either case -- a rural housewife drawing a glass of water for her child, or the farmer wheeling water over his crop -- I may be describing a threatened way of life.

Many parts of the country face severe shortages of groundwater. Some arid and semi-arid Western states are taking water from the ground faster than rainfall can replenish it. Let me mention just a few of the problem areas:

o In southern Florida, groundwater mining has so lowered the water table that in dry spells, saline water infiltrates domestic wells.

o Drinking supplies in at least one-third of the communities in Massachusetts have been affected by chemical contamination to some degree. A state report says that by September of 1979, wells had been closed in 22 communities, with losses averaging 40 percent of of the supply and ranging as high as 100 percent.

o During 1962-63, there were 150 cases of hepatitis reported in one small, rural community in Lincoln County, Montana, where almost every home had its own well and septic system. The community is on a flood plain and when the water table rises every spring, domestic sewage contaminates the wells.

o Near Denver, almost 30 square miles of a shallow aquifer were contaminated by Aldrin, Dieldrin and other toxic substances. During the 1950s, these substances had seeped from an unlined holding pond at the Rocky Mountain Arsenal, where pesticides and chemical warfare agents were being manufactured. More than 60 wells used for household supply, livestock and irrigation had to be closed down.

When my Senate committee was considering the chemical Superfund legislation, I became concerned over the impact that release of toxic chemicals was having on the environment. Since no one seemed to have any good estimate of the magnitude of the national injury, I asked the Library of Congress to review its files and furnish me with a report. Keep in mind, this was not an investigation. There was no field work; no interviews with public officials or anyone else, and no visits to distant places around the nation. It was simply a library specialist opening a file drawer and running through its contents to see what was available. It was a simple task of listing incidents that had come to the attention of one specialist -- often simply by chance. Even within this limitation, the report was further constrained. Because the Superfund law did not deal with certain poisonous substances; contamination from sewage and oil pollution were excluded.

Thus, the report was certain to be a vast understatement of the problem. Yet it documented over 1,300 wells and well fields that had been closed because of contamination.

Of those closings:

- 242 were from organic contamination;
- 26 from chloride contamination;
- 23 from nitrate contamination;
- 619 from metals contamination;
- Three from contamination by other organics;
- 185 from industrial waste disposal where the contaminant was unknown, and
- 64 from landfill leachate where the contaminant was unknown.

And, as our growing knowledge demonstrates, the threat to groundwater includes not only contamination, but exhaustion, at least in some parts of the country.

The wondrous formations of subterranean pools and rivers that lie under the surface of most of the country are a vital resource. Roughly half the population of the United States depends upon groundwater as its principal source of drinking water. About one-quarter of all water withdrawn in the nation comes from under the ground. In 1975, it was estimated that one-third of all irrigation water came from underground sources.

Some communities are totally reliant upon groundwater, while others use only surface supplies. For example, on a statewide basis, groundwater furnished only two percent of all water withdrawn in Montana in 1975; but groundwater provided 86 percent of the withdrawals in Kansas, 68 percent in Nebraska and 61 percent in Arizona.

The importance of groundwater will continue to increase. From 1950 to 1975, the total amount of water withdrawn from the ground rose from 34 billion gallons a day to 82 billion gallons a day. It is estimated that total will increase to 100 billion gallons a day by the Year 2000. Various regions of the nation will be under particular pressure, with population growth and increasing competition for limited water supplies.

Our groundwater resources constitute a vast strategic reserve, similar to the nation's huge deposits of coal. It has been estimated that thirty-six quadrillion gallons of fresh groundwater lie within a half-mile of the surface. That's more than four times the volume of the Great Lakes. It is 1,200 times the amount of water withdrawn in 1975. Note that the annual recharge of underground supplies is estimated at almost ten times the withdrawals.

But those gross figures mask some critical regional problems. In 1975, an estimated 25 percent of groundwater withdrawals were overdrafts -- that is, they exceeded recharge by natural or artificial means. Some aquifers are "fossil" in nature -- they receive no replenishment and water removed from them is said to be "mined." Many other aquifers are recharged at an extremely slow rate. All of which means that much groundwater has to be viewed as a non-renewable source.

In a few places, overdrafts have lowered the water table so much that pumping for irrigation has become too costly, and farmland has deteriorated or been taken out of production. In both coastal and inland areas, saltwater has seeped in to fill aquifers from which fresh water was overdrawn. One report I have seen says:

> "Because of the high salt content of seawater, as little as two percent of it mixed with fresh groundwater can make a portion of an aquifer unusuable under current drinking water standards."

Still another problem with overdrafting is land subsidence; the sinking of the land. This can result in structural damage, reduced property values, greater vulnerability to flooding and loss of underground storage capacity.

I recognize that you are the experts on this subject. But, I want to stress that we have a problem, and that we in the Congress are aware that it could become a big problem. We have placed an alphabet of laws on the books. One of them, the Safe Drinking Water Act, is specifically designed to seek a solution to many of the problems we have been talking about. Other laws -- Superfund, the Clean Water Act, the Resource Conservation and Recovery Act and the Stripmining Law, to name but four -- have protection of groundwater as a major purpose. Later this year, the Senate Committee on Environment will begin the process of legislative reauthorization of the Safe Drinking Water Act. The chairman of the subcommittee responsible for that law is Senator Dave Durenberger of Minnesota. He has scheduled hearings on some reauthorization issues and has plans to schedule markups early in the spring. Senator Durenberger expects to focus on the following issues:

THE DEGREE TO WHICH THE PRESENT UNDERGROUND INJECTION PROGRAM IS ADEQUATELY PROTECTING GROUNDWATER SUPPLIES.

The Environmental Protection Agency has estimated that more than eight billion gallons of hazardous waste is injected into deep wells each year. The number of wells and the volume of waste are certain to increase as federal and state regulation of landfills, surface impoundments and incinerators get tougher -- as it will. Of even greater concern, however, are the shallow wells used to inject hazardous wastes into or above aquifers. EPA estimated that, in 1978, there could have been between five thousand and ten thousand of those wells. That estimate has been dropping steadily since then. But even one such well is one too many. There is no good reason to ever inject hazardous waste into or over an aquifer that can be used for drinking water or irrigation water. Only bad things can result from that practice.

THE IMPACT OF FEDERAL REQUIREMENTS OF SMALL COMMUNITIES

It is particularly difficult for some cities and towns to finance construction of water treatment facilities needed to comply with drinking water standards. Congress has responded to this concern by allowing states to grant case-by-case extensions of up to three years to meet interim primary standards. EPA has estimated that nearly 14,000 community water systems would have been unable to meet the January 1, deadline for compliance contained in the original Safe Drinking Water Act because of inadequate lead time or lack of money.

There are some, of course, who feel the Congress should establish a National Water Utilities Bank to help establish a finance facilities. That is unlikely to happen for some time.

HOW WELL THE FEDERAL-STATE IMPLEMENTATION PROCESS IS WORKING

The Safe Drinking Water Act envisions a scheme in which states assume day-to-day responsibility for enforcing national requirements and for assuring safe and adequate supplies of drinking water. But, officials in at least one state -- Iowa -- have decided that, because of reduced federal support, the state no longer can afford to continue the primary enforcement authority for the drinking water program. While no other state has shifted responsibility for drinking water programs to the EPA, officials in several other states have discussed that possibility. A recent study by the General Accounting Office reported widespread noncompliance by local water systems, with federal drinking water standards. The consensus is that this finding reflects insufficient resources at the state and local levels and inadequate guidance from the federal level. All of these questions are tough ones, but none is new.

As fate pulls all of us deeper into the web of mounting environmental problems, I cannot help but ask whether many of our present problems haven't been intensified because we responded with easy answers and incomplete solutions in the past. The easy answer makes today's life more comfortable --but it reaps bitter fruit in the future. We have harvested our forests not just once or twice, but in some regions as many as five times. We have exhausted many mineral supplies. We have contaminated much of our surface water and air.

But our groundwaters remain relatively untouched. Groundwaters are delivered into our hands in a virtually virginal state. They are cleaner than our most sophisticated human technologies can make them. They are absolutely free. We pay not a single cent for this vast supply of pure water.

But in the past, clean air and water have been so abundant that they have not been treated like the finite natural resources they are. We now know that even though our forests are renewable, we must plant a seedling for every tree that is felled.

The same holds true for drinking water. Our future and our children's future depends on maintaining a usable natural resource. That means a resource that is safe. Not probably safe, but certainly safe.

We made a mistake with our surface waters, and we made a mistake with our air. It is costing us billions upon billions of dollars to set those mistakes right.

We cannot afford to make the same mistake with our groundwater.

CHAPTER 20

CONFERENCE SUMMATION

Timothy L. Harker, Esquire

Kadison, Pfaelzer, Woodard, Quinn & Rossi
2000 Pennsylvania Avenue, N.W.
Washington, DC 20006

The search for absolute answers to drinking water contamination problems has produced reliance on mathematical models that produce questionable risk assessment data that is taken as gospel by the press, the public, and perhaps by government (even on occasion by Congress playing to the public). If I hear the suggestion correctly, there is a vicious cycle of questionable data that generates public pressure for more regulation which, in turn, generates an increased need to rely on more questionable data. Perhaps there is a research need here to improve the risk assessment capability or, on the other hand, to educate the press and the public so as to allow regulators to escape from this vicious circle.

I heard for the first time in Dr. Cotruvo's remarks that (a) with the exception of the by-products of disinfection, drinking water normally is a minor contributor to human exposure to organic contaminants; (b) it is unlikely that drinking water will prove to be a major cause of cancer in the United States; and (c) probably other illnesses such as cardiovascular disease may prove to be more significant consequences of chemicals in drinking water.

It seems to me there are two regulatory implications of that suggestion:

1) Much government toxicology research on carcinogens through the National Cancer Institute and other institutions may be of great significance for occupational exposure, but not all that significant for drinking water programs. The question then is whether the necessary noncarcinogenic toxicology research is adequately supported by the federal government.

2) Much of the public support for the Safe Drinking Water Act is out of concern for cancer ("carcinophobia"). If that is so, and if the chlorine by-products problem ultimately is resolved, will the current need for a strong regulatory program in the safe drinking water area diminish with respect to chemical contaminants?

The "mixtures" controversy with respect to animal testing was framed as a regulatory conundrum, i.e., a conflict between the reality of exposure to chemicals in drinking water (in a mixture) and the "single-chemical" focus of environmental laws. Mr. Manwaring also touched on this issue when he rhetorically asked where the equities lie between requiring the downstream user or the upstream contaminator of the water supply to pay to clean it up. Dr. Cotruvo also suggested, as mentioned, that the cancer threat posed by organic contamination in drinking water, except for the by-products of disinfection, may not be a national hazard.

It seems to me that there are the following elements of real value for regulatory policy in each of these comments.

1) With respect to the Superfund issue of risks from hazardous waste disposal sites -- if there is a national public health problem related to abandoned sites, is it through the drinking water route? Yet, this alleged health problem may not be of national proportions (given the suggestion regarding organic contaminants in drinking water).

2) For purposes of the Safe Drinking Water Act, requiring treatment techniques is the only sane way to deal with the plethora of organics and the substantial number of inorganics. This is especially so when we don't know much about the toxicological effects of most of these contaminants. In a sense, the Safe Drinking Water Act, with its preoccupation with single compound MCLs, has stood regulatory wisdom on its head, i.e., it is better to regulate generically whole categories of compounds that are receptive to one kind of treatment technology by requiring that technology, rather than worrying about developing the toxicology and risk assessment information necessary to demonstrate that each individual compound is bad. This is especially so if we know, as we do already, that some organic compounds are harmful to health. Unlike the Safe Drinking Water Act, the Clean Water Act with its priority pollutants/technology-based approach to standards appears to take the more intelligent approach.

3) This raises Mr. Manwaring's point: in the real world, is it fair to allow a discharger under the Clean Water Act to meet technology-based standards

which allow a discharge and then turn around and require the downstream user of that water supply to treat the contaminants in the water based on a risk assessment approach? My principal thought is that in the real world, this is simply a fact of life. Without zero discharge there is no way currently to preclude discharges of contaminants to the waterways of the country by industry. It is arguably better simply to require technology-based standards for the cleanup of all drinking waters derived from surface waters and groundwater sources that contain organic and inorganic compounds. I should point out that the recent bill introduced by Congressman Eckert, H.R. 3200, would however, accomplish certain of the small system compliance issues raised by Mr. Manwaring.

On the subject of small systems compliance, the two umbrella issues are: 1) risk assesment, still at a level of reliability just beyond that of voodoo magic, as a basis for adopting MCLs that are too costly for small systems, and 2) the chronic inability of small sytems to comply, with the public health risk attendant on that fact.

I am pleased to see EPA considerations of additional flexibility in implementing the Safe Drinking Water Act, especially the emphasis on alternative sources. However, I question whether determining the availability of technology on a site-specific basis is "appropriate" within the scope of the Safe Drinking Water Act. Clearly, using the Clean Water Act as a precedent, that kind of case-by-case analysis would not be allowed under the gambit of Best Available Technology. In any event, this thinking reflects EPA's recognition of a problem that Congress perhaps has declined to deal with. EPA's attempt, administratively, to provide some relief should be applauded.

I have a question as to whether there is not significant value to be served in requiring mandatory alternatives to centralized treatment during periods of noncompliance. Recognizing that there will be more flexibility taken in the administration of the Safe Drinking Water Act with respect to small system compliance (i.e., take economics into account on a system-by-system basis, ultimately demanding compliance) this may be necessary in order to protect the public fully during the period of noncompliance.

With respect to the suggested public notification modification, I favor a notification requirement only for violations of health-related monitoring, treatment, or MCL standards. Certainly I believe that the notification should be in a nonalarming fashion. But I believe that, as an enforcement matter, there is currently too little attention paid to failure to comply with public notification requirements by small systems in circumstances that would meet my definition of when a notification should be given.

The WHO Guidelines for drinking water quality recognize that treatment technology that is <u>feasible</u> in <u>advanced</u>

countries may be impractical in developing countries and that rural areas may require different considerations from large metropolitan areas. This reference undoubtedly is to the question of microbiological control as a priority, versus placing organics on the back burner of regulation. But the logic would seem to apply in this country with respect to the current situation of noncompliance on the part of small systems (i.e., rather than tolerating continued small system violations of current standards it would be better to mandate, in the short run, alternatives to upgrading centralized treatment, pending prolonged wrestling with the question of the economic feasibility of capital improvements).

The suggestion that there is a higher consumption of drinking water per kilogram of body weight in children five years and under, considered together with the sensitivity of children to lead and nitrates is, in my judgement, a highly significant comment by Dr. Hickman. It suggests a need for special regulatory attention to systems in this country that already are on the margin or in violation of lead or nitrate standards. It suggests, also that the lead and nitrate MCLs may be too low.

The discussion by Dr. Rice on European bottled water is truly a first in this country, if not worldwide. It is of historical/cultural/sociological interest to reflect on the differences in the U.S. and European bottled water industries.

If I hear Dr. Rice correctly, the bottled water industry in Europe developed not as an alternative to tap water, but out of health or therapeutic concerns, and there is a recent restriction of that aspect, at least in the form of the EEC regulations restricting health claims for mineral waters.

In the United States (per Dr. Hutton), the historical development of the bottled water industry is the reverse, i.e., the industry initially was founded on concerns of the consumer with taste of tap waters, and it has grown more recently to include the concerns with health considerations, i.e., health concern over what is in tap water as opposed to health concern for what the constituents of the bottled water can do for one.

It is interesting to note the European requirement to state the analytical composition on the label of the mineral water product versus the U.S. FDA enforcement policy against listing of compounds on the label unless it is in compliance with the nutritional labeling standards. In my judgement, it would be very helpful to list the compounds on the label of the bottled water product without having to suggest that this somehow or other constitutes a nutritional claim. The consumer here has the priority interest.

Dr. Cotruvo referred to drinking water as a contributor of both risk and of potentially valuable trace elements. The latter aspect as highlighted, I believe, in Dr. Rice's paper on the European bottled water industry is deserving of more attention in this country.

INDEX

abietic acid 155
acetonitrile 177
acid rain 11,40
acrylamide 191
acrylamide monomer 12
acylonitrile 67,191
activated alumina 47,52,53,60
activated carbon, adsorption 99,134-139,143-145,156, 213
activated carbon, biological 134,136,139,144
activated carbon, filtration 35,36,44,45,47,54,55,57, 60,129,133,144,198,223
activated carbon, regeneration 144,146,156
additive effects 8
additives, direct 63,64,83
additives, indirect 63-66,69, 80,83
adipates 191
aerobic microbiological activity 133,139
Alachlor 191
Aldicarb 4,189,191,200
Aldicarb carbamate 49
Aldrin 3,257
algae blooms 142
algae, control with ozone 127,132,135
algal metabolism 132
alginates 14
alkanes, halogenated 3
alkenes, halogenated 3
alkylbenzenes 4
alkylphenols 151
alternative disinfectant 87, 97,98,105,106
alternative disinfection practice 105
aluminum 2,14,132,133,189,190
ammonia 22,88,89,90,131,133, 136,140,155
antagonistic effects 8
anthracene 177
antimony 65,72,76,190,252
aromatic amines 3,16
aromatic nitro compounds 3,
arsenic 2,39,44,52,53,65, 72,113,162,163,182,190, 193,210,212,252
arthritis 234
asbestos 2,12,14,33,50,54, 66,189,190,215
Atrazine 191

bacterial regrowth 129
bacteriostatic agents 244
bacteriostatic residual 128
barium 2,39,40,50,52,65,72, 163,182,190,210,212
benz(a)anthracene 177
benzene 3,4,17,164,165,187
benzene hexachloride 4
benzenes, alkyl-substituted 3
benzenes, halogenated 4
benzo-3,4-fluoranthene 252
benzo-11,12-fluoranthene 252
benzo-1,12-perylene 252
benzo(a)pyrene 177,208
benzo-3,4-pyrene 252
bryllium 2,182,190
Best Available Treatment 28, 184,218,219,265
Best Practicable Technology 184
bicarbonate 247
bioassays, animal 6

biodegradation, promotion by ozone 149,151,155
biological activated carbon 134,139
biological contaminants 190
biological processing, with ozone 133
bismuth 182
boron 182
bottled mineral water 64
bottled water 33-41,63,64, 202,209,212-220, 266
bottled water industry, European 223-254,266
breakpoint chlorination 89, 90,111,138-140,144,145, 156
bromide 96,97,98
bromobenzene 4,188
bromodichloromethane 95,97, 112,177,180
bromoform 49,96,97,177,180
Butachlor 191
by-products, of chlorination 103,105,264
by-products, of chlorine dioxide 105
by-products, of disinfectants 109,110,186,192,193,294, 263,264
by-products, of oxidation 98,99,101,125

cadium 2,6.12,14,39,40,50-52, 65,72,78,163,182,190,212, 252
calcium 10-12,76,133,182, 225,235,247
calcium bicarbonate 234
cancer 2,5-7,14,212,263, 264
cancer, bladder 11
cancer, colon 11
cancer, initiators 6
cancer, mortality rates 17
cancer, promoters 6
cancer, radiation-induced 4
cancer, rectal 11
cancer risk 14,17,264
Carbofuran 191
carbon tetrachloride 3, 12, 24,25,47-50,68,73,164, 165,180,187,188
carcinogenicity 66
carcinogenic potency 7
carcinogenic promoters 15
carcinogenic risk assessment 205
carcinogens 15,17,205,263
carcinogens, putative 6,7
carcinogens, suspect 15
carcinophobia 264
cardiovascular disease 1,10, 263
chemicals, carcinogenic 6
chemicals, mutagenic 6
chemical warfare agents 257
chloramine, di- 88,91-93
chloromine, mono- 88,91-93. 125,135
chloramines 12,87,88,93,98, 99,101,102,105,106,131, 155,193
chlorate 14,89,105,106
Chlordane 3,49,191
chloride 11,12,39,50,103, 106,151,247,251,252,258
chlorinated organics 12,138
chlorinated water 35
chlorinating agent 127
chlorination 16,95,96,98, 101,109,111-116,118,139, 142,147,185, 193
chlorination, combined residual 95
chlorination, free residual 95
chlorine 13,22,87,89,90,93, 96,100-103,106,110,113, 115,116,125,127,128,130, 134-136,138,140,142,148, 151,153,155,156,184,185, 192,223
chlorine, costs 149,150
chlorine, combined 88,91,99, 142
chlorine dioxide 13,87-89 91,92,94,95,98,99,101,102, 105,106,125,129,135,149, 151,153,155,192,193
chlorine, dose/residual 96
chlorine, free 87-89,91,95, 97-99,103,131

chlorine, organic 110,140
chlorine, residual 113,123, 131,137,141-143
chlorite 14,89,105,106
chloroacetic acid 112
chloroalkanes 206
chloroalkenes 206
chlorobenzene 164,165
chlorobenzenes 3,187,206
chlorodibromomethane 95-97, 177,180
1-chlorododecane 103
chloroform 1,3,13,16,47,48, 49,68,95-97,100,102,109-119,177,180,206
chloroform precursors 111
chloromaleic acid 112
o-chlorophenol 99
p-chlorophenol 99
chlorophenols 3,12,98,99,206
chlorosuccinic acid 112
o-chlorotoluene 4
cholera 194
chromium 39,50,52,65,72,163, 182,190,252
Clean Air Act 255
Clean Water Act 199,255,259, 264,265
coagulants 23
cobalt 182
coliform bacteria 64,163,186, 190,191,209,211,245
colony count, revivable 245
colony count, total 245
colony count, viable 244,245, 249,250,253
Community Water Supply Survey (CWSS) 164
Comprehensive, Environmental Response, Compensation and Liability Act (CERCLA) -- see also Superfund Act 198,219
copper 2,6,14,39,66,182,189, 190,251,252
copper sulfate 22
corrosion 189,190,210
corrosion control 212
costs, chlorine 149,150
costs, ozonation 145,148,150, 151
Council of the European Communities, Directive of July 15, 1980 201
cyanate 131
cyanide 2,12,131,190,206,252
cyclohexanone 67,78
cysts 56
cysts, protozoan 54

2,4-D 3,39,163,191,252
DDT 3
Dalapon 191
dechlorination, by activated carbon 138,139,250
deionization 36
dental caries 10,225,234
detergents 132
dibromochloropropane (DBCP) 40,164,188,189,191
dichloroacetic acid 101,111-113
1,2,-dichlorobenzene 4,165
1,4,-dichlorobenzene 4,49, 164,165,188
1,1-dichloroethane 188,191
1,2-dichloroethane 3,17,73, 164,165,187,188
1,1-dichloroethylene 3,164, 165,187,188
1,2-dichloroethylene(s) 3,17, 164,165,188
dichloroiodomethane 96
dichlorofumaric acid 112
dichloromaleic acid 112
dichloromalonic acid 112
dichloromethane (see also methylene chloride) 12
2,4-dichlorophenol 99
dichlorophenols 151
1,2-dichloropropane 49,188, 191
2,2-dichloropropanoic acid 112
3,3-dichloropropenoic acid 112
2,2-dichlorosuccinic acid 112,113
Dieldrin 3,257
dihaloacetonitriles 101
N,N-dimethylformamide 67,78

Dinoseb 191
Dioxin (TCDD) 191,199
Diquat 191
disinfectant, primary 87,105
disinfectant, residual 98
disinfectants 87-89,91,95,99, 101,106,125,127
disinfection 1,7,12,16,83,93, 97,109,110,123,125,126, 128,131,132,135,142,161, 184,194,211,216,224,244, 250,253
disinfection by-products 43, 87
disinfection efficiency 89,91 93
disinfection power 95,105
disinfection rate 92
distillation 36
diuresis 246
downhole sensing 170
Drinking Water Research Foundation (DWRF) 209

electromagnetic induction 170
electromagnetic resistivity 170
Emergency Water Plan, U.S. Army Corps of Engineers 220
Emergency Water Supply Plan, Department of Interior 220
Endothall 191
Endrin 4,39,50,163,191,252
environmental contaminants 184
epichlorohydrin 191
epidemiological investigations 5-7,11,14,17,23
epidemiology 7
epoxy by-products 13
ethanol 16
ethylbenzene 4,17,188,191
ethylene dibromide (EDB) 40, 164,189,191
European Economic Community (EEC), Directive of July 15,1980 201,206,242,243

Federal Food, Drug and Cosmetic Act 63
Federal Insecticide, Fungicide, and Rodenticide Act 63,199
Federal Surface Mining Control and Reclamation Act 199,259
Federal Water Pollution Control Act 220
fertilizers 1,12,13
fertilizers, nitrogen 13
fluoranthene 252
fluoridated water 36,37
fluoridation 10,23
fluoride 2,9-12,22,27,33,35, 41,44,50,162,163,186,190, 193,210,215,225,234,247, 251
fluorine 127
fluorosis, crippling 10
fluorosis, dental 10
fluorosis, skeletal 2,10
food processing 197
formaldehyde 191
Freons 191
fulvic acid, aquatic 111-113, 115,117
fulvic acids 96

gamma logging 170
gasoline, unleaded 188
gastrointestinal disturbances 6
generally available treatment technologies 64
georeferenced data sets 169
geostatistics 171
giardiasis 190
Glyphosate 191
Good Manufacturing Practices -FDA 36,38,216
gout 234
groundwater, contaminant plumes in 170
groundwater, contaminantion 161,167-169,171,173,186, 194
groundwater monitoring 198

groundwater, monitoring techniques 167-175
groundwater protection 197-200
Ground Water Supply Survey (GWSS) 25,164-166,188

halocarbons 177-180
haloforms 206,251,252
hard water 10
Health Advisories 64
health advisory levels 174
health effects 218
health risks 28,215,218,219
health threats 212
health waters(see "medicinal waters") 225,236
heart disease 6
heavy metals 12,52,54,130, 258
hepatobiliary disorders 235
Heptachlor 3,153,191
Heptachlorepoxide 3,153
herbicides 166
heterotrophic bacterial growth 57
hexachlorobenzene 4,49,191
hexachlorocyclopentadiene 191
humic acid, aquatic 104,105 111,112
humic acid, soil 99,103-105, 111
humic acids 11,47,96,97,100, 102
humic materials 131
humic materials, aquatic 105, 109,110
humic substances 12,87,95-98, 101,103,105
hydrocarbons 12,76
hydrocarbons, chlorinated 15
hydrogen peroxide 127,151
hydrogeclogical conditions 167
hydrologic cycle 9
hydroperoxide radicals 89
hydroquinone 99
hydroxyl radicals 89,127
hypertension 6
hypobromous acid (HOBr) 97
hypochlorite 13,22,87-89,91, 93,95,105,113,143
hypochlorous acid (HOCl) 87, 88,91-93,95,105,127

indene (1,2,3-cd) pyrene 252
International Bottled Water Association 35,38,217
International Sanitary Regulations 201
iodide 96
iodine 225
iron 2,14,39,95,126,130,132, 133,135,136,138,147,182, 208,235,247,250,252
isopleths, three-dimensional 171

kidney damage 2
kidney dialysis 105

lead 2,6,12,14,39,40,50,52, 65,66,72,76,163,182,189, 190,205,206,210,212,252, 266
leaking underground storage tanks (LUST) 188,189
legionella 190,191
Legionnaire's Disease 190
Lindane 3,39,50,163,191,252
lithiasis 234
lithiasis, renal 234
lithium 235

magnesium 2,10,76,77,133,182, 225,234,235,247,251
magnesium sulfate 234
malaoxon 153
Malathion 153,154
mammalian cell transformation 6
manganese 2,39,95,126,130, 135,136,138,147,182,208, 252
medicinal water's (see "health waters") 225,229,237,240, 242

Memorandum of Understanding
FDA-EPA 38,63,216
mercury 2,39,65,72,163,190, 216,252
metallic contamination 11
methemoglobinemia 2
methemoglobinemia, infantile 13
methemoglobinemia, nitrate-induced 6
Methoxychlor 3,39,50,163,191, 252
methyl butyl ketone (MBK) 67
methylene chloride 3,165,187
methyl ethyl ketone (MEK) 67, 78
methyl trichloroacetate 114
microbial contamination of water 5,7,189,211
microflocculation 133,135,139
microgeology, of contaminated sites 169
microorganisms, pathogenic 89
mineral water 224-250,266
molybdenum 182,190
monitoring techniques, remote 170
monitoring wells 170,171
monochloroacetic acid 101
mottled teeth 33
mutagenicity 6,16,17,71
mutagenic responses 16,111
mutagens 6,13,15

NTA (nitrilotriacetic acid) 11
National Contingency Plan, for responding to contamination of drinking water supplies 215
National Contingency Plan, under Federal Water Pollution Control Act 220
National Contingency Plan, under the Superfund Act 219
National Drinking Water Standards - EPA 38,70,83, 162,163,166,185,189-194, 209,210,214,217,219
national groundwater protection strategy 197, 198
National Hazardous Substance Response Plan 220
National Organics Monitoring Survey (NOMS) 95,164
National Organics Reconnaissance Survey (NORS) 95,164
National Sanitation Foundation Standard 14 54,69,70,72, 73,75,76,78
National Screening Program for Organics in Drinking Water (NSP) 164
National Water Utilities Bank 260
natural mineral water 224-225,231,242-250,253
nickel 2,182,190,252
nitrate(s) 2,6,13,27,33,39, 41,44,50,133,136,140,162, 163,190,205,210,212,215, 252,258,266
nitrification 131,133,136, 140,141,143
nitrate 131,206
nitrogen trichloride 88
nitrosamines 3

obesity 234
oleic acid 155
organic halides 100,101,106, 110,113,115,127,140,155, 156,173,226
Organic Volatile Analysis 173
organoleptic defects 245,249
organoleptic properties, of water 132,205,208
oxidant, chemical 125
oxidation potential 125,127
oxidation products, organic 128,149,153,155,156
oxidation products, partial 129,155
oxygen, atomic 127
ozonation 35,124,125,128, 129,130,132,135,139,141, 143-145,147,155,216

ozonation costs 145,148,150, 151
ozonation, three-stage 136, 143,144,156
ozonation, two-stage 136, 138,139,141-143,149
ozonator 37
ozone 13,36,87-89,91,94,95, 98,106,123-159,193,223, 250
ozone contacting 125,145
ozone contacting device 138
ozone contactor exhaust gases 138,139,146,156
ozone contact time 132
ozone disinfection 124,125, 128,129,132,133,136-141, 144,147,155,156
ozone generation 124,125,145
ozone oxidation 134,138
ozone residual 125,128,129, 136-138
ozone transfer efficiency 136,137,146

polyelectrolytes 14,22
polyethylene 172,182
polyethylene, chlorinated 77
polyhydroxyaromatic oxidation products 151
polynuclear aromatic hydrocarbons (PAHs) 2,3,12, 16,177,191,208,251,252
polyvinyl chloride (PVC) 66, 67,73,77,78,172
polyvinyl chloride, chlorinated 66
post-chlorination 137,142, 144,145
Post-closure Liability Trust Fund 174
post-ozonation 137,141,142, 145
potassium 182,235
prechlorination 134,138,140, 142,144
preozonation 132,134,136-145, 155,156
priority pollutants 177,178, 264

PVC piping 15
paraoxon 153
parasites 211,245
Parathion 153,154
pathogenic organisms 94,226, 245,249
pentachlorophenol 4,191
perhydroxyl radicals 127
permanganate 130,151,153
pesticides 1,3,12,13,24,33, 50,64,132,153,162,163, 166,167,172,173,185,186, 189,191,199,200,206,210, 216,226,251,252,257
pesticides, halogenated 172
phenols 24,39,49,72,78,99, 132,138,151,178,252
phenols, chlorinated 12,24
phthalate esters 3,191
Picloram 191
piezometric head gradients 171
point-of-use units 43,45,46, 48,49,53-57,60,61
polychlorinated biphenyls (PCBs) 3,50,173,191,199

Quality Standard - FDA 36, 38,216

radiation 210
radioactivity 33,39,216
radionuclides 2,4,64,162, 163,185,186,192,194,204,
radium 27,193
radium-226 4,163,192
radium-228 4,163,192
radon 4,192
Recommended Maximum Impurity Content (RMIC) 84
regionalization 219
remedial actions, costs of 171
renal dialysis 204
renal infections 234
residual vinyl chloride monomer (RVCM) 72,76
resistivity logging 170
Resource Conservation and Recovery Act (RCRA) 28,168,173,199,259

reverse osmosis 36,37,44,45, 50,60
reverse osmosis-activated carbon units 43,46,47, 50,55
Rural Water Survey (RWS) 164

Safe Drinking Water Act 28, 38,63,64,83,161,162,173, 185,193,199,200,209,214-218,220,259,260,264,265
selenium 2,27,33,39,65,72, 162,163,182,190,193,215, 252
silicates, activated 14
silver 2,12,39,50,52,58,59, 163,182,190,206,252
silver, deposited on carbon 57
Simazine 191
small water systems 26-28, 30
sodium 6,12,41,50,143,182, 189,190,212,215,234,235
sodium chloride 225
sodium chlorite 22,89
sodium-restricted diets 33
soft drinks, mineral water use in 229
soft water 10
spring water(s) 226,230, 231,235,236,250
Standard Reference Materials, National Bureau of Standards 177
strontium 182
styrene 191
succinic acid 112
Suggested No Adverse Response Levels (SNARLs) 64
sulfate 190,225,247,251,252
sulfide 131
sulfur 250
sulfur compounds 244
Superfund Act 28,33,168,198-200,219,220,257-259
synergistic effects 8
synthetic organic chemicals 186,189,191,194,212

2,4,5-T 3
2,4,5-TP Silvex 39,163,191, 252
table water 230,233,235,237
Teflon 172
terpineol, alpha 155
tetrachloroethylene 3,17,49, 66,164,165,177,180,186-188
tetrahydrofuran (THF) 67,78
thallium 182,190
tin 72,76,78
tin-based heat stabilizers 68
toluene 4,17,188,191
Toxaphene 4,39,163,191,252
toxicological studies 16,23
Toxic Substances Control Act 28,63,199
triaryl/alkyl phosphates 17
triazine herbicides 4
trichloroacetaldehyde 112
trichloroacetic acid 101, 109,111-118
trichlorobenzene(s) 4,164, 191
1,1,1-trichloroethane 3,49, 73,164,165,187,188
trichloroethylene 3,24,40, 47,49,73,164,165,177,180, 186-188,191,213
2,4,6-trichlorophenol 4
2,3,3-trichloropropenoic acid 112
trihalomethane formation 101, 110,147
trihalomethane precursors 126,134,142
trihalomethanes (THMs) 1,12, 13,14,17,22,39,47,49,50, 64,72,73,87,88,95,97-100, 102,103,105,106,109,110, 126,134,141,142,144,147, 162,163,177,185,186,192, 210,212,214,216,251,252
typhoid 194

underground injection control program 199,200,259
uranium 4,192

urinary infections 234

vanadium 182,190
vegetation, decay products of 1,2
vinyl chloride 3,12,15,68,165, 187,188
viral hepatitis 235
virucidal efficiency 91,93, 94
viruses 191,211
virus inactivation, with ozone 128,147
Volatile Organic Chemicals (VOCs) 24,26,45,47,64, 162,164,165,166,172,177, 186,187,192,211,212
Vydate 191

wastewater reuse 23
Water Treatment Chemicals Codex 14,64,83,85
water systems 43,51
World Health Organization (WHO), Environmental Health Criteria Programme 203
WHO, European Drinking Water Standards 202
WHO, Guidelines for Drinking Water Quality 201,203
WHO, guideline values 203-206,208
WHO, International Standards for Drinking Water, (1958) 201,202

xylenes 17,191
m-xylene 4,188
o+p-xylene 188

zinc 2,12,14,39,76,182,189, 190,252